Studies in Space Policy

Volume 35

Series Editor
European Space Policy Institute, Vienna, Austria

The use of outer space is of growing strategic and technological relevance. The development of robotic exploration to distant planets and bodies across the solar system, as well as pioneering human space exploration in earth orbit and of the moon, paved the way for ambitious long-term space exploration. Today, space exploration goes far beyond a merely technological endeavour, as its further development will have a tremendous social, cultural and economic impact. Space activities are entering an era in which contributions of the humanities — history, philosophy, anthropology —, the arts, and the social sciences — political science, economics, law — will become crucial for the future of space exploration. Space policy thus will gain in visibility and relevance. The series Studies in Space Policy shall become the European reference compilation edited by the leading institute in the field, the European Space Policy Institute. It will contain both monographs and collections dealing with their subjects in a transdisciplinary way.

The volumes of the series are single-blind peer-reviewed.

Marco Aliberti · Ottorino Cappelli · Rodrigo Praino

Power, State and Space

Conceptualizing, Measuring and Comparing Space Actors

ESPI
European Space
Policy Institute

Springer

Marco Aliberti
European Space Policy Institute
Wien, Austria

Ottorino Cappelli
Dipartimento di Scienze Umane e Sociali
Università di Napoli L'Orientale
Napoli, Italy

Rodrigo Praino
College of Business, Government and Law
Flinders University
Adelaide, SA, Australia

ISSN 1868-5307 ISSN 1868-5315 (electronic)
Studies in Space Policy
ISBN 978-3-031-32870-1 ISBN 978-3-031-32871-8 (eBook)
https://doi.org/10.1007/978-3-031-32871-8

This Springer imprint is published by the registered company Springer Nature Switzerland AG
The registered company address is: Gewerbestrasse 11, 6330 Cham, Switzerland

Foreword

The ESA Agenda 2025 clearly spelled out Europe's ambition to assert itself as a *space power* in the same league as the US, China and Russia by 2035, the 60th anniversary of ESA creation. Admittedly, the continent already stands second to none when it comes to leveraging space for fundamental science and downstream applications in the areas of telecommunications, Earth observation, meteorology and navigation. But for most of its space journey, it has never exploited the full strategic dimension of space, nor has it invested in leadership autonomy across the full spectrum of activities. This is the challenge that confront us today.

As for any other journey, to properly guide our pathway in this grand journey, there is a need to understand where we currently stand and what the requirements to reach the destination are, while in the process keeping track of every progress and possible drawbacks.

This new Springer book in the series "Studies in Space Policy" is a most welcome contribution in this regard, as it provides an innovative and thought-provoking way to put in perspective Europe's current place in the international space arena and understand the possible trajectories than can bring the continent in the league of space powers.

Beyond contributing to scholarly debate on the concepts of *spacepower* and *space power*, the book provides a valuable toolkit to inform decisions that Europe and other space actors need to make to tackle the challenges ahead. With its rich sets of indicators devoted to measuring and classifying space actors in a multidimensional context, the book clearly shows where action is required to effectively fulfil Europe's space power potential.

The multidimensional approach to spacepower explained in this book is consistent with the priorities set forth in the ESA Agenda 2025, which posits that Europe's potential in space can only be unlocked by advancing European leadership in many key different key domains, from space exploration to secure communications, passing though autonomous and competitive access to space and commercialization efforts.

Starting with the field of human and robotic exploration, Europe needs not only to remain a partner of choice for major international cooperation initiatives; it also needs establish an independent capability for undertaking its human spaceflight missions.

As the recent report "Revolution Space" made it crystal clear, there is much at stake. A bold exploration programme "would galvanise and revolutionise the whole European economy, well beyond the space sector, and inspire a generation of Europeans to build the future."

Europe also needs robust, sustained investment in cutting-edge and world-class space science missions and R&D efforts for ground-breaking space technologies that will put it at the forefront of global innovation dynamics. The SOLARIS preparatory initiative for space-based solar power is a poignant example of these ground-breaking advances. By the same token, Europe needs to maintain and further consolidate its existing leadership in Earth Observation and Meteorology (with the Copernicus Sentinel Expansion Missions and Sentinel New Generation) as well as GNSS (with the Galileo second generation and LEO-based PNT). All European stakeholders are also called to join all their forces in the development of IRIS^2 for ensuring secure communications for public and private use and capturing a share of this growing market.

Equally fundamental, Europe needs to ensure a fast inaugural flight of Ariane 6 and the return to service of Vega-C while paving the ground for a completely new, more competitive and reusable launcher system so that European autonomy in accessing space will be preserved. It similarly needs to be the forefront of the international sustainability efforts in space and of the flowering market for in-orbit servicing activities. While undertaking cutting-edge space operations, it also needs to strengthen the safety the space activities, particularly in the area of space weather monitoring and operational service provision, asteroid deflection, and debris collision avoidance services.

More broadly, Europe needs to boost its efforts in space commercialization and ensure our industry's ability to capture a larger share of the booming global space market.

Those are the crucial tasks of the present decade. Time is of essence. If Europe wants to grow as a global leader in space, alongside the US, Russia, China and other prominent space partners by the mid-2030s, it needs to take the right steps and decisions today. To echo the recent admonition of the High-Level Advisory Group on Human and Robotic Space Exploration for Europe: "Europe needs to recognize that this entails prompt action to catch-up and leapfrog the competition, and shape the future in line with Europe's values. The cost of inaction would far outweigh the necessary investment to establish Europe as a strong and independent space actor."

Thanks to its legacy, expertise, knowhow and competitive industrial capacity, Europe *can* become a space power. Crucially, it *must* become a space power, for it cannot afford to be left behind the unfolding "space revolution," with its high multiplier effects across the economy and society. But real political will and a bold vison at the highest political level is required. Every great achievement—be it in space or elsewhere—requires such a vision and political commitment. The analysis presented in this book takes us one step closer to the realization of this formidable task by pointing out in detail our strengths, our weaknesses, and some specific ways we can improve our position in relation to other major space actors.

ESA is ready for the challenge and committed to play its role in shaping and implementing Europe's ambition in space—an ambition of inspiration, exploration and realization of Europe's political, societal and economic goals. It now behoves European statespeople at the highest political level to come together and jointly forge a grand European vision that can take space and Europe to the next level.

Paris

Josef Aschbacher
ESA Director General

Acknowledgements

The authors would like to thank the many people and organizations that made this book possible.

On a financial note, this research was supported by the Australian Government through a grant by Defence. The views expressed herein are those of the authors and are not necessarily those of the Australian Government or Defence. The grant was competitively won and administered by SPPARC, the Space Power and Policy Applied Research Consortium. Hosted by the College of Business, Government and Law at Flinders University in Adelaide, Australia, SPPARC brings together in a vibrant research consortium Flinders University's extensive expertise in space policy and space law with the well-established expertise in this area of the European Space Policy Institute in Vienna, Austria, and the traditional expertise in Political Science of the University of Naples "L'Orientale" in Naples, Italy. Researchers from all three institutions were actively engaged in this international research collaboration for three years.

On a more academic note, the authors discussed some of the ideas presented in this book at the 2021 Annual Conference of the European Consortium of Political Research (ECPR, August 30–September 3) in a panel chaired by Rodrigo Praino and entitled "State, Society and the Space." The authors would like to thank the other participants to the panel for their insightful comments, including Luciana Cingolani, Hertie School, Berlin, Andrea Molle, Chapman University, and Andrea Vaccaro, Sapienza University of Rome.

The authors would like to express sincere gratitude to Matteo Cappella and Francesco Maria Villani for their continuous and most valuable support throughout the course of this project. Without their precious dedication and most valuable assistance in the collection of data and the execution of the comparative assessments presented in this study, this book would not exist in its current form. The assistance of Ph.D. candidate Eva Palo in collecting the relevant political science literature was also much appreciated.

More broadly, the authors wish to express great appreciation to the entire ESPI team for the collection of hard data, and in particular to Hermann Ludwig Moeller,

Mathieu Bataille, Yui Nakama and Ferechta Paiwand for their useful inputs and feedback.

Furthermore, the authors would like to highlight the invaluable assistance offered by the befriended colleagues at the University of Naples "L'Orientale" and Flinders University, Vinicius Guedes Gonçalves de Oliveira in particular, whose critical suggestions further contributed to refine the quality of the research.

The authors are finally grateful to all experts and stakeholders interviewed under Chatham House Rule during the preparation of this book and to the many people that reviewed it and provided constructive feedback. Sincere thanks are in this context extended to the many anonymous country experts that completed the "Measuring Spacepower" survey underpinning the expert-coded assessments presented in the study. Although we need to preserve their anonymity, we want to express our heartfelt gratitude, confident that they will recognize themselves.

Of course, all errors, shortcomings, and inaccuracies remain the sole responsibility of the authors.

Contents

Chapter 1
Introduction

> India today registered itself as a space power. Till now, three countries of the world - America, Russia, and China had this achievement. India is the 4th country to have achieved this feat. Narendra Modi (2019)

> Europe is a significant space power: its space sector had a turnover of €74 billion in 2019, which represents between 15 and 20% of the world market. Joseph Borrell (2022)

> To explore the vast cosmos, develop the space industry and build China into a space power is our eternal dream. Xi Jinping (2022)

> I will free NASA from the restriction of serving primarily as a logistics agency for low-Earth orbit activity—big deal. Instead, we will refocus its mission on space exploration. Under a Trump Administration, Florida and America will lead the way into the stars. Donald Trump (2016)

During the height of the Cold War, being a space power was virtually a requirement for any nation aspiring to superpower status. At the time, possession of a manned space programme was probably enough to claim space ower status. That status was perhaps more symbolic than substantive, but it was fundamental nonetheless. Today, space and space-related activities have become a much larger substantive part of everyday life, with space technology assisting a wide range of activities in an ever-growing number of disparate domains. It was just a matter of time before this substantive increase in importance was accompanied by a number of symbolic milestones. In fact, after a long hiatus, states are once again actively engaging in space activities that showcase the ever-growing importance of space in the twenty-first century and generate overarching implications stretching across various domains including political, commercial, and technological competition. The establishment of the United States Space Force in late 2019, the planting of the Chinese flag on the surface of the Moon in December 2020, and the first crewed launch undertaken by a private company in May 2020 are just a few recent landmark developments. The upcoming return of humans to the earth's satellite, which will certainly include the first woman to ever land on it, as well as the ambition to take humans all the way to Mars for the first time are some of the symbolic breakthroughs that are expected to occur in the next decade or two.

M. Aliberti et al., *Power, State and Space*, Studies in Space Policy 35,
https://doi.org/10.1007/978-3-031-32871-8_1

While more and more countries start their own national space programmes, claiming or re-claiming membership in the international "space club" (Paikowsky, 2017), what makes a country a space power remains largely a matter of opinion. Soon after the nationally acclaimed and internationally condemned anti-satellite test conducted by India in March 2019, for instance, Indian Prime Minister Narendra Modi affirmed that India had become an elite space power thanks to the successful execution of the test (Reuters, 2019). On a similar vein, before the publication of the 2022 white paper on space activities, President Xi Jinping was reported to convey China's perpetual yearning to venture into the cosmos and establish a robust space industry, positioning the nation as a fully-fledged space power (CNSA, 2022). During the 2016 US presidential campaign Donald Trump pledged to ensure that America would "lead the way into the stars" (Cillizza, 2020), and in Europe, soon after the release of the 2022 Strategic Compass, the High Representative of the European Union for Foreign Affairs and Security Policy Joseph Borrel asserted that Europe holds the status of a significant space power due to its substantial market share in the global space industry (Borrell, 2022).

In spite of the momentum gained by the body of research on space policy and politics in the past decade, the scholarly community has yet to achieve a consensus on what *spacepower* is and which countries are eligible to *space power* status. Arguably, this difficulty largely stems from the nuanced, intrinsically political and contested nature of the concept of "power." Power is indeed a challenging concept, as evidenced by the panoply of different, often conflicting conceptualisations found in academic literature across various disciplines and sub-disciplines.[1] What is also critical is the unresolved duality between power as a *material attribute* (a set of means and resources) and power as a *status*—the actual quality of being powerful, which does not simply stem from the number of soldiers, money, bombs or satellites a nation may possess.

Thus, despite the apparent self-explanatory nature of the concept, there still are a number of outstanding questions regarding its inner meaning and constituent elements. What are the defining features of power *in* space? Does the acquisition of any measure of *spacepower* automatically grant a country the status of *space power*? On what basis can a nation claim such status? And what are the criteria differentiating a *space power* from "lesser" space actors? How should these differences be taken into account? And perhaps most importantly, how can *spacepower* be empirically measured and countries endowed with *space power* status identified?

To date, there is no efficient and comprehensive way to conceptualise, measure and compare spacepower. Scholars, practitioners and politicians alike frequently delve into discussions surrounding the concepts of *spacepower* and *space power*, but to our knowledge not only there is no consensus, but most of the proposed conceptualisations lack the necessary foundation for measurement, let alone for performing a thorough comparative analysis.

---

[1] As a point of reference see the *Encyclopaedia of Power* edited by Dowding (2011), which contains 381 entries written by 157 authors from 17 countries.

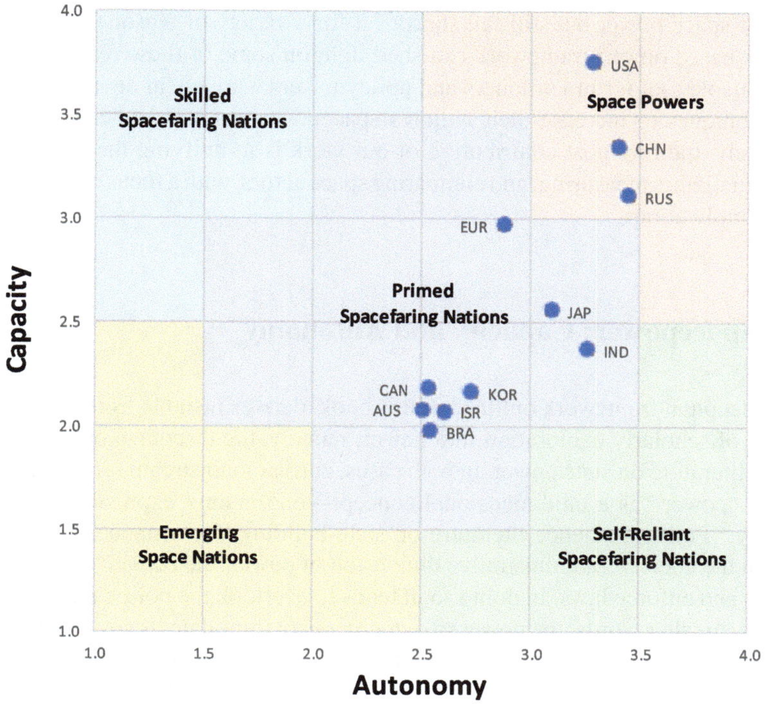

Fig. 1.1 The spacepower matrix

In this book we propose an original conceptualisation of spacepower that takes into account the interaction between two distinct dimensions: decision-making autonomy and executive capacity. We then employ this novel conceptual framework to construct a measurement model that allows us to operationalise the concept by measuring spacepower across both the capacity dimension and the autonomy dimension. Finally, we collect empirical data for 11 space actors (i.e., Australia, Brazil, Canada, China, Europe, India, Israel, Japan, South Korea, Russia, and the United States) and draft a comparative analysis of the spacepower of each one of these actors. Ultimately, we are able to locate the position of each of the actors considered within a bidimensional matrix that we call the *spacepower matrix* (see Fig. 1.1).

Once empirical data are analysed, only the United States, China, and Russia emerge as full-fledged space powers. But even within this very exclusive club of nations there are marked differences that can be empirically measured. In fact, while the United States emerges as the most capable of the three space powers, China and Russia present higher levels of autonomy. In essence, the three space powers excel in different ways and, consequently, are each better positioned to deal with different space-related issues. The remaining 8 space actors considered are what we propose to call “Primed Spacefaring Nations.” In other words, they are space actors with great potential that, to varying degrees, can and should aspire to achieving the status of space power but still fall short of it for a variety of reasons. An empirical analysis based on our framework can shed light on some of these reasons, and can be a valuable tool for practitioners and policymakers who aim at devising strategies and techniques to increase their nation's spacepower vis-à-vis other space actors. Ultimately, the foremost contribution of our work is in unifying the three tasks of conceptualizing, measuring, and comparing space actors, with a focus on the resulting policy implications.

1.1 Spacepower: Capacity and Autonomy

The conceptual framework outlined in this book derives insights from two different spheres of scholarly exploration into power: namely, the discourse on space power and the literature on state power. In both cases, current mainstream scholarship tends to treat "power" as a unidimensional concept—one mainly expressed in terms of *capacity*.[2] Political science literature on state-building, for instance, focuses basically on the coercive-administrative dimension of power, the *capacity* to implement policies and enforce laws. In doing so, it tends to overlook the political dimension of the state, in other words, the power to *autonomously* formulate its sovereign will and articulate it into public policy decisions independently from or even against the will of divergent interests (Barkey & Parikh, 1991: 525). Political autonomy vis-à-vis both foreign and domestic powers, however, is pivotal to define modern state authority.[3] A government entity capable to implement policy decisions taken elsewhere would not qualify as a "state" by modern standard. To acquire the status of "state power" it would need to possess a quantum of political autonomy as well. State power is indeed a multidimensional concept and "stateness" is a measure composed of two equally crucial dimensions: political autonomy and administrative capacity.

Having set this as our starting point, we contend that the literature on space power presents limitations strikingly similar to those found in studies on state power. Conceptually, it is overwhelmingly capacity-oriented and largely overlooks the fundamental issue of political autonomy in space matters. Methodologically, it lacks a clear, shared consensus on what is to be measured and how to assess the outcome of the exercise in a comparative perspective. Operationally, its ability to offer useful information and guidance to policy makers is limited.

This is due first and foremost to the neglect of the distinction between *spacepower* and *space power*, a distinction that was introduced three decades ago by David Lupton (1988) but was not coherently followed in subsequent studies. Elaborating on Lupton, we consider *spacepower* as a quali-quantitative measure based

---

[2] The conceptual framework presented here was first developed by political scientist Ottorino Cappelli as a heuristic tool for the comparative study of state building (Cappelli, 2010). It was subsequently adapted and operationalised by space policy analyst Marco Aliberti to serve as a guide to the conceptualisation and comparative measurement of spacepower (Aliberti et al., 2019).

[3] The literature on state autonomy is abundant. As a point of reference see the seminal work of Michael Mann (1984). For a broader overview see Chapter 2.

on interplay between two basic dimensions: the autonomous decision-making power of a state in space-related matters, and its material capacity to implement *its own* policy decisions and achieve the related goals in the political, diplomatic, military, economic, or social realms. Furthermore, we propose that autonomy and capacity be further divided into hard and soft subdimensions. The hard subdimensions can be measured by objective data and include *hard capacity*, which encompasses material assets and skills enabling operations across the spectrum of space activities, and *hard (or technical) autonomy*, which evokes the notion of self-sufficiency and pertains to resources and expertise necessary to ensure mission independence in case of necessity. The soft subdimensions, on the other hand, have been measured by administering a special survey to a poll of experts. They include *soft capacity*, which refers to the adept utilisation and integration of space assets and expertise within national policies, infrastructure, and activities, and *soft autonomy*, or political autonomy proper, which signifies a state's ability to devise national space strategies and policies, independent from or against the will of divergent interests, be them foreign or domestic. Ultimately, the combination of all these dimensions and subdimensions results in a multi-fold typology of space actors where only those possessing high levels of both capacity and autonomy may claim the status of *space powers*. The other actors are labelled *emerging space nations* and *spacefaring nations* with the latter category being further subdivided into *skilled space nations*, *self-reliant space nations*, and *primed space nations*.

Arguably, one of the innovative features of our framework lies in the introduction of the autonomy dimension. Autonomy in fact—both in its "hard" or technical and its "soft" or political form—is the inescapable *sine qua non* for explaining power as status, in sociology and political science as in the field of space studies: no centre or power can be fully understood without addressing the fundamental question of whether it is able to independently determine its own interests and formulate strategic decisions on how to pursue them. In this way our study fills both a theoretical and an empirical gap in the current understanding of *spacepower* and *space power*. Theoretically, it sets forth the idea of spacepower being a multidimensional concept resulting from the complex intertwining between "hard" and "soft" capacity and "technical" and "political" autonomy. Empirically, it lays the foundation for collecting and processing a large amount of hard-to-find data on spacepower and for aggregating multiple indicators into a comprehensive measurement model and an in-depth comparative assessment encompassing the most active space actors worldwide. This allows us to measure *spacepower* by means of both objective ("observable") and subjective ("expert-coded") data as well as to employ these data to identify *space powers* and differentiate them from other types of space actors.

The underlying goal is to enable scholars, space professionals and decision-makers of different countries to understand how to empirically evaluate and strategically enhance a country's relative position vis-à-vis other actors engaged in space activities. Indeed, by applying the aforementioned framework, it becomes possible to understand what the requirements are for a country to become a *space power*. Specifically, this book has been designed to significantly increase both the quality and the quantity of the discourse on space policy and strategy, providing insights into how space nations can increase their relative status as a major strategic actor in space, hence developing, projecting, and sustaining national capabilities in a way that can shape their position and role in the field.

Last but not least, the book aims to support evidence-based decision-making. Consistently, it is in the authors' hope that the information, data, and materials included in the research will prove of help to those designing policy strategies and advicing policymaking entities.

1.2 Book Outline

This book is articulated into three main parts. After this introductory overview, Chap. 2 lays down the general conceptual framework for measuring spacepower and comparing space actors. Building on the assumption that spacepower is a form of state power, the chapter starts with a critical review of the scholarly literature on the state and articulates the concept of state power into two constituent dimensions, namely that of *capacity* and *autonomy*. It shows that the overstretching "tyranny" of the concept of state capacity—which heavily leans towards implementation and quantifiable, hard data—often overshadows the inherently political notion of autonomy, which relates to the decision-making side of the policy process and is considerably more challenging to assess empirically. Yet a political entity capable to execute decisions that it has not autonomously taken would not qualify as "a state," however powerful it may be or appear. The chapter then turns to the literature on space power, which presents limitations strikingly similar to those found in studies on state power. Conceptually, it predominantly focuses on capacity and largely neglects the fundamental issue of political autonomy in space matters. Methodologically, it lacks a clear, shared consensus on what is to be measured and how to assess the outcome of the exercise in a comparative perspective. Operationally, its ability to offer useful information and guidance to policy makers is limited. To correct this situation, the chapter sets the twofold goals of reframing the concept of capacity and of developing a way to assess—conceptually as well as empirically—the autonomy of the state in the realm of space policy. The chapter concludes that only actors possessing considerable levels of both autonomy and capacity in space matters (i.e., high levels of spacepower) can be considered *space powers*.

Chapter 3 offers a methodology to operationalise the conceptual framework described. Specifically, it provides a detailed description of how spacepower can be disaggregated into its two constituent dimensions of capacity and autonomy, and how these can be further broken down into the four subdimensions of hard capacity, soft capacity, hard (or technical) autonomy and soft (or political) autonomy. For each of these subdimensions, the chapter features an in-depth description of the constituent indicators and provides a detailed explanation of the underlying measurement model. Specifically, scores for the indicators of the *hard* subdimensions of capacity and autonomy are based on quantitative data retrieved from a number of specialised sources, while the indicators for the *soft* subdimensions result from qualitative evaluations coming from leading space power experts around the world who compiled an ad hoc survey designed and administered for this study. These data have then been combined into a composite index of *spacepower* and are visualised in the form of a matrix featuring a non-hierarchical typology of space actors.

Chapter 4 discusses the empirical results of the research. It focuses on the comparison of different space actors along the two prominent dimensions of capacity and autonomy and the related subdimensions. The countries selected have been among the most active in space in the past decade and have attracted the most attention from general media outlets, practitioners, and scholars alike. They are Australia, Brazil, Canada, China, India, Israel, Japan, Russia, South Korea, United States and Europe, taken as a whole. Building on the results of the measurement exercise, the chapter provides a comparative assessment of the status and relative positions held by the analysed actors within a multidimensional and non-hierarchical matrix of spacepower based on the two dimensions described above. The matrix shows the relative position of each space actor at a given point in time and helps interpret their trajectories throughout a multi-fold typology of space actors that includes: (1) *space powers* (fully autonomous and capable actors), (2) *skilled spacefaring actors* (fairly capable, but insufficiently autonomous actors), (3) *self-reliant spacefaring actors* (reasonably autonomous, but insufficiently capable), (4) *primed spacefaring actors* (fairly capable and autonomous) and (5) *emerging space actors* (that are yet to attain a satisfactory level of both autonomy and capacity).

Chapter 5 synthesises all the major findings of the study, offers some additional reflections, and indicates some avenues for future research.

With its original classification and in-depth comparisons based on an extensive corpus of hard-to-find data, it is our aspiration that this book will equipe scholars, space professionals and policymakers with a valuable analytical toolkit. This toolkit will aid in scrutinizing and contextualizing different countries' strategic ambitions in space, along with the necessary means and resources to accomplish these objectives.

References

Aliberti, M., Cappella, M., & Hrozensky, T. (2019). *Measuring Space Power: A Theoretical and Empirical Investigation on Europe*. Springer.

Barkey, K., & Parikh, S. (1991). Comparative Perspectives on the State. *International Review of Sociology., 17*(2), 523–549.

Borrell, J. (2022). *Space and defence: protecting Europe and strengthening our capacity to act*. European External Action Service. https://www.eeas.europa.eu/eeas/space-and-defence-protecting-europe-and-strengthening-our-capacity-act_en. Accessed on 30 August 2022.

Cappelli, O. (2010). Pre-Modern State-Building in Post-Soviet Russia. In Hill, R. and Cappelli, O. (Eds). *Putin and Putinism*. Routledge.

Cillizza, C. (2020, 27 May). 12 truly amazing Donald Trump quotes on space. *CNN*. https://edition.cnn.com/2020/05/27/politics/trump-spacex-rocket-astronauts-elon-musk/index.html (Accessed on 20 August 2022).

CNSA (2022). *China's Space Program: A 2021 Perspective.* The State Council Information Office of the People's Republic of China. China National Space Administration. http://www.cnsa.gov.cn/english/n6465645/n6465648/c6813088/content.html (Accessed on 20 August 2022).

Dowding, K. M. (ed.). (2011). *Encyclopaedia of Power.* SAGE.

Lupton, D. E. (1988). *On Space Warfare: A Space Power Doctrine.* Air University Press.

Mann, M. (1984). The Autonomous Power of the State: Its Origins, Mechanisms and Results. *European Journal of Sociology, 25*, 185–213.

Paikowsky, D. (2017). *The Power of the Space Club.* Cambridge University Press.

Reuters (2019, March 27). *Modi Hails India as Military Space Power after Anti-Satellite Missile Test.* https://www.reuters.com/article/us-india-satellite-idUSKCN1R80IA (Accessed on 20 August 2022).

Chapter 2
Conceptualizing Space Actors: State and Power in Space

This chapter sets forth a comprehensive, cross-disciplinary conceptual framework aimed at a dynamic comparison of space actors, taking into account their technical, industrial and political characteristics as well as the trajectories that bring them in or take them out of 'the club' of space powers. It contends that space power is a form of state power. It also argues that to acquire the status of space power, a state must possess a certain degree of *spacepower*, a multidimensional concept comprising two main attributes: capacity and autonomy. *Autonomy* refers to the state's ability to formulate space-related interests of its own and devise national space strategies independent from or against the will of divergent interests both foreign and domestic. *Capacity* relates to the state's ability to pursue those interests and implement its strategies in the political, diplomatic, military, economic, or social realms. The combination of these two dimensions results in a multi-fold typology of space actors, which consists of *space powers*, *emerging space nations*, and *spacefaring nations*. The latter group is then further divided into three distinct subgroups: skilled spacefaring nations, self-reliant spacefaring nations, and primed spacefaring nations.

2.1 Space Power as a Form of State Power

Over the past two decades, the post-Cold War talk of a presumed "end of history" (Fukuyama, 1992) has been swiftly replaced by an ever-fiercer competition among great (and lesser) powers. In parallel, the struggle to explore, exploit and control outer space has become more and more relevant to governments, the military and the business community alike. Consistent with these transformations, the concept of *space power* has grown increasingly popular in the scholarly debate, military doctrines and public policies around the world. Despite the apparent self-explanatory nature of the concept, however, there is no universally accepted definition of *space power*, let alone a comprehensive theory explaining its actual meaning and implications. As

M. Aliberti et al., *Power, State and Space*, Studies in Space Policy 35,
https://doi.org/10.1007/978-3-031-32871-8_2

a consequence, several issues surrounding the concept remain largely unresolved, and the key features that define a country as a space power are currently unclear, making it virtually impossible to properly differentiate between space powers and lesser space actors or determine how a country can even become a *power* in the space realm. This difficulty largely stems from the fact that the very concept of "power" is slippery, as shown by the panoply of different, often conflicting conceptualisations found in political science, sociology and international relations literature (Dowding, 2011). What is essential, for our purposes, is the unresolved duality between power as a *material attribute* (a set of means and resources) and power as a *status*—referring to the actual state of being powerful.

Admittedly, early space power literature presented a promising perspective on this issue when D. E. Lupton (1988: 4) proposed to differentiate between *spacepower* ("the ability of a nation to exploit the space environment in pursuit of national goals") and *space power*, a term denoting "a nation [endowed] with such capabilities." But this line of reasoning got somehow lost as subsequent authors generally used the two terms indistinctively to indicate both space actors and the set of space capabilities that such actors may possess—although in varying degrees. Furthermore, Lupton himself did not conceptually clarify what exactly the essential *components* of spacepower are and what role they play in the making of a space power. Specifically, he left the question open as to what is needed—beyond mere "strength"—to turn even a capable space actor into a full-fledged space power.

With the aim of putting some order into such a complex matter, in this chapter we will revert to Lupton's use and speak of *spacepower* when referring to the attributes necessary for an actor to be deemed a *space power*. We will focus on how to distinguish space *powers* from lesser (less powerful) space actors—or, in other words, on how to *measure* spacepower and *define* the different statuses that derive from such measurement. In this connection we argue that spacepower should not be treated as a simple, unidimensional concept focused primarily on the material *capacity* of a state to *pursue* specific policy goals in the space realm. We suggest instead a more complex conceptualisation of spacepower that also incorporates the crucial political dimension of *autonomy*—i.e., the ability of a state to *decide* what goals it wants to pursue. Our use of spacepower thus upgrades Lupton's in that it identifies two different though strictly interrelated components, or dimensions—namely decision-making autonomy and implementation capacity. It also stresses that states (or supranational entities as in the case of Europe)[1] must possess a certain degree of both autonomy and capacity to acquire the status of space power. Other combinations, such as strong capacities but weak autonomy (or the reverse), would lead to different statuses, none of which is coincident with the actual quality of being a full-grown, *powerful* space nation.

Last but not least, our framework is based on a parallel between space power and state power. In both cases, in fact, mainstream literature tends to treat "power" mainly in terms of capacity. Most works on state building, for instance, focus almost

---

[1] Europe is considered here as a single space actor. Considerations on this matter are provided in Chap. 4.

exclusively on how to build effective administrative structures capable to execute policies and enforce laws. In doing so, they tend to overlook the political dimension of autonomy, i.e., the sovereign power of the state to formulate national goals and articulate them into policy decisions "independent from or even against the will of divergent interests" (Barkey & Parikh, 1991: 525). However, autonomy vis-à-vis both foreign and domestic powers is pivotal to define modern state authority.[2] A governing entity capable to implement policy decisions taken elsewhere would not qualify as a "state" by any modern standard. State power in other words, is a multidimensional concept and any attempt to define and measure it should take into account the interaction between the two equally crucial components of political autonomy and executive capacity. The same, we argue, applies to the concept of space power, which we characterise as a form of state power. We proceed from the assumption that only nation states have the means, incentives and legitimacy to master the entire spectrum of power—including first and foremost the power to independently define strategic national interest, make binding decisions about it, and act authoritatively to implement them. Thus, although non-state, private actors may be endowed with some of the material attributes of power in the space policy realm, only states can achieve the levels of both decisional autonomy and executive capacity that allow an actor to claim the status of space power.

Arguably, one of the innovative features of our framework lies in the introduction of the autonomy dimension. Autonomy, as we define it, encompasses both a "hard" or technical aspect and a "soft" or political dimension,[3] and it is the inescapable *sine qua non* for explaining power as status, in sociology and political science as in the field of space studies: no power can be fully understood without addressing the fundamental question of whether and to what extent it is able to independently determine its own interests and formulate strategic decisions on how to pursue them.

2.2 State Power: Capacity and Autonomy

One of the most striking features of the contemporary literature on the state is the conceptual confusion that surrounds the notion of state power. We argue that this is due, in good part, to the conflation of two fundamental dimensions of *stateness* that have had divergent fortunes lately: state capacity and state autonomy.

State capacity, which has been by far the most studied dimension in the past three decades, pertains to a set of material and organisational resources and skills that enable states to enforce laws, implement policies and achieve goals. As the next section shows, the overstretching and ultimate "tyranny" of the concept of capacity—overwhelmingly focused as it is on the state's excecutive capabilities and measurable, at least in part, by hard data—has obscured the inherently political and sometimes intractable concept of the *autonomy* of the state, which is closely related

[2] For a review of the relevant literature see Sects. 2.2 and 2.3 in this chapter.

[3] For the distinction between "hard" and "soft" autonomy see below, Chaps. 3 and 4.

to the notion of sovereignty. Yet a state with limited capacity of implementation and enforcement may be dubbed a weak or failing state, but an entity that is capable to execute decisions which it did not make, is not a "state" by any modern standard. Neglecting this factor only reflects the "destatisation" mood of the past decades. Indeed, for all the rhetoric of "state building" that has been spreading lately, what is generally forgotten is "stateness" itself, a notion that has political, decision making autonomy at its core. This is indeed what distinguishes "modern states"—regardless of the type of regime—from what we call 'vassal' or client states internationally, and 'captured' states domestically. In other words, in order to really "bring the state back in," *state power* must be conceptually understood and empirically assessed as a multidimensional status pertaining to political-administrative entities that possess some substantial and possibly measurable degrees of both decisional autonomy and executive capacity.

2.2.1 *State Capacity: The Tyranny of a Concept*

It is commonly recognised that the concept of state capacity has a tendency to conflate different dimensions of the state and stretch across different domains incorporating an array of "closely related attributes of states such as strength, fragility, failure, effectiveness, efficiency, quality, legitimacy, autonomy, scope." As a result, it is surrounded by conceptual confusion and "continues to be a concept that lacks precise definition and measurement" (Hanson & Seigman, 2021: 2; Cingolani, 2013, 2018; Vaccaro, 2022). Such a misformed and overstretched concept (Sartori, 1970) is also "tyrannical," in its attempt to encompass the entirety of *stateness* and be equated with state power *tout court*. This occurs at the expense of another crucial dimension of state power that is erroneously incorporated into the concept of capacity and somehow "buried" within it, namely state autonomy. To conduct a rigorous evaluation and measurement of stateness and state power, it is necessary to distinguish between capacity and autonomy and then examine their interrelations.

2.2.1.1 A Misformed, Overstretching Concept

State capacity is measured by two different types of indicators, those based on objective hard data drawn from various statistical sources, and those based on soft data, i.e., compiled from surveys or expert assessment.[4]When using the first set of

[4] Objective indicators used to measure state capacity produce two sets of measures. The first set comprises (a) macroeconomic measures like total GDP and per capita income, (b) statistics regarding public services and social welfare, such as secondary education, public housing, and healthcare facilities, among others, and (c) data related to the extent and quality of infrastructure, including postal services, telecommunications, road and highway networks, etc. The second set of objective indicators focuses on a state's ability to perform essential governmental functions, such as (d) protecting borders and waging war, as measured by factors like military spending and the percentage

indicators several difficulties arise from these being strongly related to levels of wealth, which leads to common-sense objections such as "Why would we expect a middle-income state to have the same level of highways as the wealthy, no matter the quality of governance?" More substantive objections may also be raised. It has been observed, in fact, that "what appear to be relatively effective states may simply be rich enough as to allow an element of inefficacy while still performing at acceptable levels;" on the other hand, poorer economies may be less likely to generate top-level welfare services and infrastructures "regardless of the capacity of the state to regulate them" (Enriquez & Centeno, 2012: 143). In sum, the critical point is that these indicators are not (or not only) measuring state capacity; they are on to something else, and this complicates both definitional and measurement tasks.

Regarding the indicators that address the fundamental aspects of stateness, such as military strength, law enforcement, and taxation capabilities, there are also numerous issues of interpretation. For instance, the extent and reach of conscription are closely linked to the geopolitical context, and nations with high proportions of their population in the armed forces are frequently in a state of perpetual war preparation, either for offensive or defensive purposes (Enriquez & Centeno, 2012: 144). This raises paradoxical questions such as whether military-controlled states with compulsory mass conscription, like Egypt, should be seen as having greater capacity to wage war or defend their territory compared to civilian-controlled states with smaller, professional armies and a voluntary recruitment policy, like the US. Concerning "Law and Order" indicators, how would data on crime and homicide rates relate to indicators evaluating human rights and the rule of law, which are also frequently utilised by various measures of state capacity? Would a police state capable of violently but effectively repressing crime be considered more capable than a liberal state imposing stronger constraints on police activity *and* experiencing higher crime rates? Lastly, regarding extraction capacity, hard data on taxation tell us little about when taxes should be considered "too high" in relation to the GDP. Hence one may wonder whether a state that levies a greater portion of GDP in taxes should be deemed just more capable than one that collects a lower percentage, or if the former is actually engaging in over-extraction or even outright predation. Furthermore, a decrease in tax revenues may be due to a decline in the state's fiscal extraction capacity, but it could also depend on the ascent to power of a political party with a robust tax-reduction agenda. This, in turn, could have detrimental effects on the future ability of the state to provide quality services, welfare and infrastructure. In such a case, data on both the level of taxation and the provision of public services would be addressing not an issue of capacity, but one of political priorities. Dozens of analogous instances might be identified by examining how researchers attempt to employ hard data to measure a slippery conceptual variable such as state capacity.

Faced with this type of challenges, analysts and scholars often turn to surveys or qualitative expert assessments, commonly referred to as "soft data." This approach

of the population serving in the armed forces, (e) maintaining law and order, measured by the size and budget of police forces, as well as crime and homicide rates, and (f) collecting taxes, measured by metrics like tax revenue as a percentage of GDP.

allows for a more careful conceptual exploration of the variables, clarifying to the respondents what each indicator aims to assess, and thereby obtaining more thoughtfully-considered responses to objectively intricate inquiries. However, this approach can also give rise to broader conceptual and methodological issues allowing us to better identify issues of concept misformation and overstretching, bringing to light what we call the "tyrannical" nature of the concept of state capacity.

To begin with, some of these indicators are overly vague in their formulation as well as in what they promise to deliver. One of the indicators in the World Bank's Government Effectiveness Index purports to measure "the quality of policy formulation and implementation." Quite a formidable task to fit into one indicator (Kaufmann et al., 2010). Additionally, several indicators are based on faulty assumptions about state capacity and how comparative measures should be constructed. Some of these are of overtly normative nature and pertain to the type of regime rather than the capacity of the state. In these cases, the underlying definitions often reflect specific features of democratic regimes, making it hard to include many non-western nations in a comparative framework. For example, some indicators offered by the World Bank, such as Voice and Accountability and Rule of Law are actually measures of liberal democracy. The same is true for two of the indicators included in the State Fragility Index (FSI) published by the Fund for Peace: State Legitimacy, which measures people's perception of the government's "openness and representativeness," and Human Rights and Rule of Law, which endeavours to identify, among other things, the presence of "authoritarian rule." These indicators convey the misleading impression that a non-representative, authoritarian regime is inherently associated with a low-capacity state. Although its curators underline that the index "does not necessarily make a judgment on democratic governance," they do just that rather than measuring state capacity (Fund for Peace, 2022). Now, as argued by Enriquez and Centeno (2012: 136), "capacity is not and should not be a normative concept"—that a state uses its capacity to pursue "good" or "bad" ends "does not negate the empirical reality of that state's ability." Samuel Huntington (1968: 2) expressed this concept vividly half a century ago, when he wrote that "The United States, Great Britain, and the Soviet Union have different forms of government, but in all three systems the government governs … in many [other] cases, governments simply do not govern." Huntington's interpretation of state capacity ("a government that governs") is admirably straightforward: the probability that the decisions of the political authorities "will be implemented through the government machinery."

The specular opposite to the normative approach based on Western standards is embedded in indicators that are tailored to address the unique circumstances found in non-western nations. For example, the aforementioned State Fragility Index includes an indicator called "Monopoly on the Use of Force" which aims to assess, among other things, whether "the military is under civilian control" (Fund for Peace, 2022). This is an important aspect of *stateness* in all countries and one that is inexplicably little considered in most indices, thus the effort to tackle this issue by comparing 179 nations sounds promising. However, the dichotomic language employed in describing the indicator fails to capture the more nuanced ways in which the military may influence civilian leaders in Western societies, involving cultural factors, organisational interests, and bureaucratic negotiations.[5] Consequently, countries in the West tend to be clustered at the top of the scale, with limited possibilities for further differentiation.

A most serious problem with these measures is the lack of clarity regarding the constituent elements of state capacity, especially when evaluated in different historical or cultural context. Measurements of bureaucratic quality and corruption may illustrate this point.

Concerning bureaucracy, the World Bank's Government Effectiveness Index has an indicator measuring the civil service's "independence from political pressures." The Quality of Government Index developed at the University of Gothenburg also presents an indicator of "bureaucracy quality" aiming to assess whether there is an established mechanism for the "recruitment and training of public officials," and whether the civil service "tends to be somewhat autonomous from political pressure"[6] (Teorell et al., 2020: 367). Basically, these indicators assume as essential components of state capacity a technically skilled *and* politically independent bureaucracy, which they present as the basic pillars of the "Weberian" concept of bureaucracy.

This seems to ignore that any political elite—including elected, legitimate decision makers—need ensure bureaucratic *loyalty*. As Theda Skocpol (1985: 16) puts it, the availability of "skilled *and* loyal officials" is one of the basic underpinnings of state power. Max Weber (1994 [1919]) himself described how an essential feature of party government is taking a share in the direction and disposition of the offices of the state, either by the distribution of offices among party adherents or by exerting pressure in other ways. The ancient concept of bureaucrats serving "at the pleasure of" the king, president, or mayor is closely related to the need of political elites to ensure that those individuals are loyal to them and will carry out their decisions in a timely and effective manner. The power to remove civil servants at any time also serves as a mechanism of control over the bureaucracy, allowing political elites to replace those who obstruct their decisions or do not adhere to their directives. This was after all the classical justification for the *spoils system* in Andrew Jackson's America in the nineteenth century (Fukuyama, 2014: 109). Throughout the twentieth century, in the US and other Western nations, the notion that the political loyalty of the bureaucracy should be ensured through party patronage has been progressively (although not completely) replaced by the idea of a competent and skilled body of officials who prioritise adherence to laws and procedures over political affiliation. But in several countries that have not experienced a comparable trajectory of political development, the dominant approach for political elites to secure "a government that governs" is still by controlling the spoils of government via personal ties and political connections. To summarise, a politically dependent bureaucracy does not necessarily indicate a lack of state capacity. On the contrary, it may serve to ensure, in Huntington's terms, that political decisions "will be implemented through the government machinery."

---

[5] In the next section, we delve briefly on the issue of civil-military relations and how they impact on state autonomy, rather than capacity.

[6] The QoG Index assesses the institutional strength and quality of the bureaucracy with a focus on whether, when government changes, the bureaucracy has the *strength and expertise* "to govern without drastic changes in policy or interruptions in government services."

A somewhat similar point can be raised about measures of corruption and related notions such as patrimonialism and clientelism, which are widely employed in studies of state capacity. Most organisations involved in creating state capacity indices deal with such topic, including the Quality of Government Institute and the Varieties of Democracy Institute, both affiliated with the University of Gothenburg (Coppedge et al., 2020). The most widely used measure is Transparency International's Corruption Perceptions Index, or CPI (Transparency International, 2022). But, like in the case of bureaucratic autonomy, there are serious problems in equating corruption with a lack of state capacity. First and foremost, the basic assumption may be untrue or at best overstated. Research shows that corruption weakens state capacity when it is used to "buy disobedience," inducing officials to violate not only the law but also the expectations of higher-ups in the administrative hierarchy. But in other situations corruption may be *informally institutionalised* and "systematically tracked, monitored, and granted by state leaders as an informal payment in exchange for compliance." This form of centralised corruption (or "state-strengthening graft") can be essential to reinforce hierarchy and provide the foundation for bureaucratic *obedience*. It may even reflect the fulfilment of an informal "contract" between state leaders and their subordinates in which graft plays a critical role (Darden, 2008: 41). This line of argumentation goes a long way towards explaining why countries like China, Russia or India, which have comparatively high corruption rates, also demonstrate a definite capacity to govern and achieve high levels of growth and political stability. State capacity measures that give for granted the state-weakening impact of corruption are unable to capture these situations.

Second, and most significant for our topic, some authors distinguish between types of corruption according to the dimension of state power they affect. According to studies that will be reviewed later (see Sect. 2.3), administrative or "petty" corruption affects state *capacity*, intended as the ability to ensure law enforcement and policy implementation. On the other hand, "grand" corruption or "state capture" impacts the *autonomy* of the state, as what is commercialised—and effectively privatised—is the essence of public decision-making, including legislative, regulative, and judicial prerogatives (Hellmann & Kaufmann, 2000). Again, the analytical distinction between capacity and autonomy is crucial to conceptualise *and* measure state power as well as state weakness.

The latter line of reasoning brings us to a final observation concerning what we consider the core problem of state capacity's conceptual over-stretching, namely the incorporation and burial of the concept of state autonomy. What is crucial here is the differentiation between decision and implementation. Comparative measures of state capacity reviewed so far tend in fact to suggest that all states "want to deliver" (and want it with the same intensity) while different rates of success or failure depend on the availability of material resources, the quality of the state's institutional design, the organisational structure or the technical-managerial competence of state personnel—in one phrase: on the state's capacity of implementation and enforcement. Is this assumption correct? And what about the state's *decision* to deliver? Is it a reflection of the political and ideological priorities of the state elites? But to what extent do the latter enjoy the political independence needed to make these decisions, let alone the capacity to actually implement them? And to what extent can they translate their priorities into coherent legislation, regulations and operative programmes, in the face of potential opposition from divergent interests? Taken from a different angle: Do divergent interests exist that may oppose state decisions and outcomes that do not align with their goals, and how do they operate? Do they limit themselves to target state capacity, distorting the *implementation* of policies that are already in place or the *enforcement* of existing regulations that they see as potentially damaging? Or, alternatively, do divergent interests target the *decision-making* process itself, promoting policies of their liking and the adoption of laws and regulations that favour them? What means do they use to these ends and how can we measure the effectiveness of their actions? Finally, the question arises: where do these divergent interests originate? Are they generated in society (typically in the private business sector), or do they emerge within state institutions (such as the bureaucracy, the military, or state-owned enterprises), or else are they located externally (like other states, international financial institutions, or multinational corporations)?

These questions regarding the decision-making power of the state (or lack thereof) should be asked separately from those regarding its executive capabilities, avoiding hastily measuring state power solely in terms of outcomes. To achieve this, one must initially separate the two interrelated components of state power conceptually, and then develop distinct empirical measures for each. Eventually, these measures can be combined in such a way that "state power" is determined by their intersection. Such is the approach that we adopt in this book.

2.2.1.2 Reassessing State Autonomy

A Short History of State Autonomy and Capacity

In the early period of state formation in modern Europe, establishing sovereignty domestically and internationally was the main challenge. As English philosopher Thomas Hobbes (2017 [1651]) stated in the *Leviathan,* "auctoritas facit legem": the laws are not derived from any natural or divine source, but rather from the authority of the sovereign, without which they are merely "words" with no power to bind or compel. Successful state formation thus depended on the ability of early state makers to make war, secure order, and collect taxes—relying on a gradually acquired monopoly of violence as necessary (Tilly, 1975). Such was the foundation for a politically *autonomous* and administratively *capable* public authority endowed with the power to *make* and *implement* binding decisions. This entity aimed at a high level of centralisation and insulation from society. Although the modern state was hardly a deliberate creation, few would doubt that political autonomy was an essential precondition of the new power structure that began to emerge in Europe around the sixteenth century (Finer, 1997). There followed a long and violent process of state making.

By the nineteenth century, however, two different strands of political though—liberalism and Marxism—had emerged that embodied powerful anti-statist ideas.

Liberals viewed the state's autonomy and insulation from society as a major threat that could lead to despotism and focused on "taming the prince" through constitutionalism, the rule of law, and accountability. As fears of a resurging Leviathan proved justified at the dawn of the new century by the rise of totalitarian dictatorships, the subsequent defeat of fascism in World War II propelled the anti-statist sentiments of liberalism to coalesce with democratic ideals as the guiding ideology of the "free world." As a result, the notion of the autonomous power of the state became unpopular in mainstream liberal-democratic thought. At about the same time, Marxism too rejected the notion of an autonomous state separate from society, albeit from a different angle. According to this school of thought, the state is merely an instrument of class dominance. In capitalist societies it serves the bourgeoisie, in a socialist society it will be a tool for establishing the dictatorship of the proletariat (after the Russian revolution, the Bolsheviks acted on this principle). However, Marxists maintained that as a result of the socialist revolution and the abolition of classes, the state would lose its *raison d'etre* and would gradually "wither away." In sum, both Marxists and liberals considered capacity and autonomy as distinct elements of state power. But, although both viewed the state as a usable concentration of coercive authority and administrative capabilities, state power could never (for the Marxists) or should never (for the liberals) be autonomous from the forces and interests generated in the civil society.

Since the 1960s, the previously normative approach of liberals began to evolve into an ostensibly realistic portrayal of the functioning of liberal democracies. Rejecting the idea of the state as an independent political actor, advocates of interest group liberalism or "pluralism" introduced in its stead the concept of a political system, a "black box" that converts societal demands and pressures (inputs) into governmental decisions (outputs). The "state"—to the extent that its existence was recognised at all—was assimilated into a broader arena where values are negotiated and resources are distributed. This left limited scope for exploring state power in its own right, conceptually as well as empirically. Concurrently, a new wave of neo-Marxist ideas emerged that recognised some degree of autonomy in states that were mediating between capital and labour through welfare and redistribution policies. However, such autonomy was seen as "relative" and temporary, arising from a balance in class conflicts. The state remained a capitalist state, with its political elite and administrative machinery serving, in the final analysis, the interests of the dominant class. The Marxist revaluation of the state, thus, proved to be half-hearted—and short-lived.

Ironically, the latter part of the twentieth century saw anti-statist ideas from the right and the left converge laying the groundwork for neoliberalism, a philosophy that prioritises society over the state and champions liberalisation, deregulation, and privatisation. Following the fall of the Soviet Leviathan and the end of the Cold War, the globalisation movement emerged, further propelling the state-reducing agenda of neoliberalism to the forefront. This culminated in the *destatisation* agenda of the so-called "Washington Consensus." It wasn't until the late 1990s that the need of having effective states regained momentum, prompted by the failures of post-communist privatisation and the global financial crises that originated in Asia. These factors underscored the importance of institutions capable of regulating the market, safeguarding property and contracts, and fostering development. Thus, the so-called "post-Washington consensus" was formed with *state building* as its mantra.

But in contrast to the historical experiences of modern Europe, contemporary state builders exclusively focus on state *capacity*, with no interest in establishing a new Leviathan that could exert sovereign *auctoritas* domestically or internationally. The state to be built is merely an instrument for policy implemention, lacking the inherent will and political autonomy necessary to intervene in the policy formation process. As a result, state capacity metrics mainly evaluate policy outcomes while ignoring or underestimating political factors related to decision-making. In a global order that prioritises economic performance over political sovereignty, inputs are fixed and originate from external sources beyond the control of the state. This approach has led to the concept of capacity becoming "tyrannical," with the ability of a state to implement exogenous decisions being mistaken for the fundamental quality of being a state. Hence, research on the state and its power remains confined in a conceptual iron cage.

Bringing State Autonomy Back In

In the 1980s and 1990s, while neoliberalism and globalisation were starting to spread, a most serious attempt was made by scholars of the *new-statist* school to "bring the state back in." Fresh new studies of the state emerged and a rich debate developed "below the radar" where proponents and critics of new statism gave their best. In that period notions of state autonomy (or lack of it) regained momentum and received fruitful conceptualisation. Then over two decades followed where all this was virtually silenced by mainstream, quantitative analysis of state capacity in terms of policy outcomes. This is why we turn to studies of that period today, to seek inspiration in challenging the tyranny of the conceptual iron cage imposed by the notion of state capacity. As these studies demonstrate, capacity and autonomy are separate dimensions of state power and state power itself can only be measured as the intersection of these two dimensions.

The manifesto of the new-statist school is *Bringing the state back in*, a collection of essays published in 1985 with the goal of analysing the mechanisms by which the autonomous power of the state operates (Evans et al., 1985). In the introduction essay, Theda Skocpol defines *autonomy* as the ability of a state "to formulate and pursue goals that are not simply reflective of the demands or interests of social groups, classes or society." In turn, *capacity* refers to the ability to "implement official goals, especially over the actual or potential opposition of powerful social groups or in the face of recalcitrant socioeconomic circumstances" (1985: 9). Thus, state-society relations are essential in both definitions, as the state is responsible for providing collective goods that cannot be provided "by partial interests." This requires the state to operate on a broader and more inclusive perspective than is feasible for "private actors embedded in the market." Civil society, on the other hand, hosts within itself strong and divergent forces that are "bent on capturing parts of the state apparatus and using them for their purposes" (Rueschemeyer & Evans, 1985: 68, 56). Therefore, establishing autonomy from these forces and purposes is regarded as crucial for any public authority involved not just in state-building, but in building stateness.

The rich and extensive discussion surrounding this topic can be condensed by briefly examining three authors who have presented three distinct but interconnected models for conceptualizing and evaluating stateness, all of which emphasise autonomy as their conceptual core.

The 'Stateness Continuum'

Joel Migdal's book *Strong Societies and Weak States* is dedicated to explaining the failure of modernisation in Africa and Asia. It criticises the "illusion" shared by "hopeful modernizing leaders" as well as advocates of state-centred approaches, namely that the state may be used to shape society like "a chisel in the hands of the new sculptors." This illusion entails taking for granted what should instead be open to question, i.e., "the issue of the autonomy and strength of the state" (Migdal, 1988: 4, xvi). The subject of the study thus is "state capabilities or their lack" as seen in a relational framework that pits the state *against* society. In fact Migdal's definition of *state strength* is predicated on the ability of state officials to utilise state institutions "to get people in the society to do what they want them to do." Conversely, *societal strength* is defined as the ability of "some social elements [to] oppose state leaders successfully, sometimes to reach unexpected accommodations with state officials, and even at times to capture parts of the state" (Migdal, 1988: 9). Society's ability to resist directly challenges the autonomy of the state. Migdal quotes Badie and Birnbaum (1983: 35) to underscore that state autonomy only manifests as "a tangible reality" when the state has the ability "to act on its environment and to autonomously impose collective goals distinct from the private goals generated within the social system itself." The opposite situation occurs when state leaders prove "unable or unwilling" to overcome social resistance to their designs (Migdal, 1988: 9).

This approach based on state-society (or public–private) confrontation suggested going one step further than what was being done by the new statists who, Migdal (1988: xvi) contends, either parachuted European notions of state autonomy onto non-Western countries or dismissed the state entirely "talking of *stateless* or *nonstate* societies." Migdal uses instead the concept of "stateness" as a *continuum* that ranges from weak to strong states (and vice versa, from strong to weak societies). Quoting Metin Heper's synthesis of J. P. Nettl's original concept, Migdal (1988: 260) emphasises that stateness is a matter of degree and varies across different polities or historical periods, "depending upon the extent to which the major goals for society are designated and safeguarded by the state, independent of civil society" (Heper, 1985: 86). Figure 2.1 visualises Migdal's simple model of the stateness continuum.

Fig. 2.1 Joel Migdal's 'Stateness Continuum'. *Source* authors' own elaboration

Despite some potential confusion in terminology,[7] Migdal's underlying message is clear: states and societies are in constant conflict: while a strong state can "displace or harness" competing powers in order to mould and shape society, in a strong society these powers can undermine the state by resisting its commands or even "making rules against the wishes and goals of state leaders."[8] Notwithstanding its extremely dichotomous approach, this vision is useful to start organizing the notions of state capacity and autonomy in accordance with our interests. Last but not least, using a continuum to illustrate the varying degrees of stateness helps to place the variable within a measurement-oriented framework that is conceptually not biased towards society or the state, but looks at state-society relationships with a realistic empirical eye.

The 'Stateness Triangle'

Another study of relevance here is *Embedded Autonomy*, an influential book exploring the quest for the most effective approach to creating a developmental state. Author Peter Evans (1995: 12) likened the state to a set of apparatuses that can function according to distinct ideal–typical models, with the two most prevalent models representing the opposite poles in a continuum of state-society relations. At one end is the "Weberian bureaucracy" model characterised by selective recruitment and career patterns resulting in cohesive and coherent structures that are effectively *insulated* from society. This extreme and potentially despotic form of state autonomy, that roughly corresponds to Migdal's *strong state*, is supposedly associated with high governmental capacity. In reality, the opposite is true as an insulated state has limited ability to gain intelligence on society and involve external, private actors in governance processes. On the other end of the continuum is a model consisting of extremely permeable state structures totally *immersed* in a network of social ties that are essentially personal connections between powerful self-interested individuals, rather than connections between social constituencies and the state as an organisation. When this is the only source of cohesion, individual maximisation takes precedence over the pursuit of collective goals, severely restricting the state's ability "of transcending the

[7] As Migdal does not provide a clear conceptual differentiation between state capacity and state autonomy, there may be some ambiguity in his terminology. He employs the term "capabilities," which appears to be interchangeable with "strength," a concept that at times encompasses autonomy, while in other points of the book the "autonomy and strength of the state" are treated as distinct elements.

[8] These competing social powers are listed as "families, clans, multinational corporations, domestic enterprises, tribes, patron-client dyads" (Migdal 1988: 31).

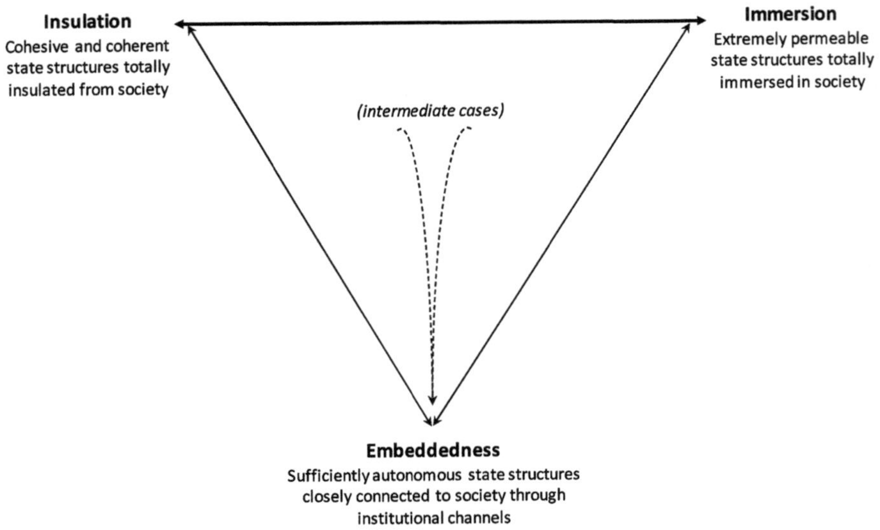

Fig. 2.2 Peter Evans' 'Stateness Triangle'. *Source* authors' own elaboration

individual interests of its private counterpart" (Evans, 1995: 12). This is reminiscent of the *strong society* depicted by Migdal.

To counterbalance these two extremes, Evans introduces the concept of "embedded autonomy," which combines the state's internal, bureaucratic cohesion and its external, social connectedness. In this model, state structures possess sufficiently autonomous commitment and coherence while being "embedded in a concrete set of social ties that binds the state to society and provides institutionalised channels for the continual negotiation and renegotiation of goals and policies." Evans argues that only when embeddedness and autonomy are combined can a state be considered "developmental" and be capable of successfully participating in, and steering industrial transformation. Nevertheless, the author concludes that very few states, such as South Korea, have administrative systems that closely resemble this ideal type. Most are considered instead "intermediate cases with partial and imperfect approximations of embedded autonomy," with Brazil and India being in this category (Evans, 1995: 12).

Based on the latter passage, if cases that fall in between are not considered examples of embedded autonomy, then *embeddedness* does not occupy a middle ground on a scale. Instead, it represents a third pole in a triangular diagram—a "political elsewhere" with respect to the simple continuum that ranges from the state's total *immersion* into society to its complete *insulation* from it. This is a welcome modification of Migdal's basic linear model, as it allows for a more refined assessment of the conceptual intricacies and factual variations of state power forms. Figure 2.2 visualises our interpretation of Evan's model.

The 'Stateness Matrix'

In a seminal article that anticipated his four-volume series *The Sources of Social Power*, British-born sociologist Michael Mann systematised the crucial issue of "the autonomous power of the state" (Mann, 1984). He provided a distinction between two dimensions of state power that he calls "despotic" and "infrastructural" and that closely approaches a distinction between the (decisional) autonomy and the (executive) capacity of the state. *Infrastructural power* is "the capacity of the state [despotic or not][9] to actually penetrate civil society, and to implement logistically political decisions throughout the realm." *Despotic power* refers to the "range of actions that the ruling elite can take without engaging in routine, institutionalised negotiation with civil society groups" (Mann, 1984: 188–189).

On this basis Mann built a double entry table or "matrix" of stateness (see Fig. 2.3) which is a major source of inspiration for our work. The infrastructural power dimension is roughly equivalent to the concept of state capacity as we intend it. Regarding the other dimension, Mann's definition is perhaps too extreme. His characterisation of it as *despotic* power where decisions are made by state elites *without* negotiation with society adds unnecessary stress to the model. Perhaps, it would be more suitable to call it "decisional autonomy" and reserve "despotic" for the highest point in a continuum where varying degrees of autonomy might be qualified by different terms including "high," "medium," "low," and so on. In our own adaptation of Mann's matrix we proceed along this line, arguing—with Rueschemeyer and Evans (1985: 61)—that "state autonomy is a prerequisite for effective state action."[10]

		Infrastructural power	
		low	*high*
Despotic power	*low*	Feudal	Bureaucratic/Democratic
	high	Absolutist/Patrimonial	Authoritarian

Fig. 2.3 Michael Mann's 'Stateness Matrix'. *Source* adapted from Mann (1984a: 191; 1993: 60[11])

[9] The phrase "despotic or not" is added to the definition in the second volume of *The Sources of Social Power* (Mann, 1993: 59).

[10] Although there is no need to delve into the details of Mann's matrix, which only serves here as a methodological inspiration, one point is worth considering. The typology presented suggests that the "bureaucratic-democratic" state has high infrastructural power but low despotic power. This is because this type of state "is controlled by others, civil society groups, but *their decisions* once taken are enforceable through the state's infrastructure" (Mann, 1984: 191). This might be taken to imply that the bureaucratic-democratic state does function as a "tool" in the hands of society, which is where real decisional power rests—as both pluralists and Marxists argue concerning the link between interest groups, social classes, and the state. This aligns with the literature on "state capture" and lobbying in the United States (as discussed below in Sect. 2.3) as well as with the outcomes of our survey, which indicate that in the field of space policy, the US displays a remarkably high capacity index while scoring the lowest in the autonomy index (see Chap. 4).

[11] Note that in the 1984 article the title of the columns in the table was 'Infrastructural Coordination', and became 'Infrastructural Power' in the 1993 book. In addition, in 1984 Mann used the term

Despite these critical points, Mann's conceptualisation offers several valuable insights. Firstly, it shares the conflictual view of state-society relations which we find, with different emphasis, in the works of Migdal, Evans and the new statist school. Secondly, it highlights that decision making autonomy and implementation capacity are two separate but interrelated dimensions of state power, an aspect that will be lost in subsequent conceptualisations of state capacity as an all-encompassing notion. Thirdly, Mann's framework underscores the need for "measuring" these dimensions by situating each of them on its own continuum ranging from low to high. This marks a step forward in comparison to both Migdal's linear and Evans' triangular models, as it rationalises the space in which the two dimensions intersect by dividing it into four quadrants, each possessing its own typological definition.

In our conceptual framework we consider all of the methodological recommendations that have arisen from the studies examined thus far. These include the suggestion that the degree of variation in state-society relations can contribute to the measurement of stateness in different countries (Migdal, 1988; Nettl, 1968), the notion of assessing individual cases based on their spatial position, rather than a linear scale (Evans, 1995), and the proposal that distinct typologies of state power arise from the intersection of its two fundamental dimensions (Mann, 1993).[12] However, our approach is both more ambitious and narrower in scope compared with those presented above. On the one hand, it purports to expand beyond the confines of state-society interactions. While we do analyse societal forces, with a particular emphasis on economic actors, we also consider two "non-societal" actors capable of influencing or capturing public decision-making: one is situated outside the polity and comprises foreign states, the other operates within the state apparatus itself, namely the domestic military. Therefore, our understanding of stateness incorporates not only state-society relations but also international relations and civil-military relations (details on this will be elaborated below). On the other hand, our approach is narrower in scope as our main focus lies in measuring stateness and assessing state power. Consequently, a comprehensive examination of the foundations of social power is beyond the scope of our analysis. Our attention towards societal and other agents of influence is residual, meaning we are primarily interested in them only to the extent that they impose constraints on the autonomous power of the state.

"Imperial" to define the quadrant at the intersection between high despotic power and low infrastrtuctural power. However, he specified that the power configuration captured therein "corresponds to the term patrimonial state used by writers like Weber and Bendix." Then in 1993 Mann used the couple "Imperial/absolutist" for the same quadrant. We conflated these definition as "Absolutist/Patrimonial."

[12] We have made a deliberate choice not to include in our analytical review the otherwise captivating narrative presented by Daron Acemoglu and James Robinson in *The Narrow Corridor* (2020). In their work, the authors construct a matrix that juxtaposes the "power of the state" and the "power of society" as the axes. They argue that when these two forces pull with the same force, a country can enter or remain in the corridor—a middle ground that allows for the attainment, maintenance, or advancement of liberty. Essentially, this entails adopting Joel Migdal's dichotomous state/society approach and representing it in the form of a matrix. Aligned with the authors' research objective, this primarily functions as a visual representation of the concept of liberty rather than as a tool for conceptualizing and measuring stateness.

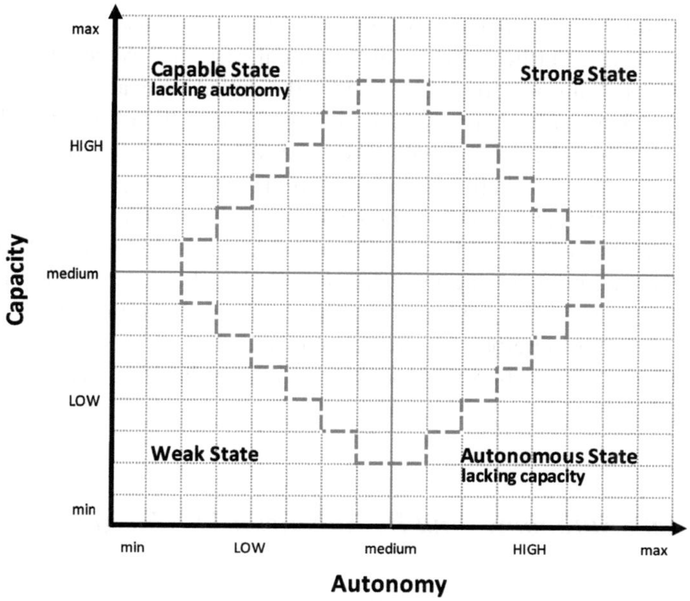

Fig. 2.4 The capacity/autonomy matrix of stateness. *Source* authors' own elaboration

Building on the foregoing, we develop a framework that untangles the concept of autonomy from the tyranny of capacity and allows us to conceptualise autonomy and capacity as distinct yet interrelated dimensions of stateness. Then we construct a comparative metric that can be applied to different countries charted across a two-dimensional plane. In this way we can overlay measurements onto the typological quadrants of a matrix and visualise each state's position relative to others at a given point in time. This is crucial because, while belonging to a particular typology provides valuable conceptual information for a state, understanding *where* it is situated *within* a quadrant provides precious analytical information that should not be overlooked. Our approach is illustrated in Fig. 2.4.

As in the matrix developed by Mann, low/high autonomy and low/high capacity intersect to create macroareas, or quadrants with associated typologies. However, rather than forcing different realities into rigid conceptual typologies, our approach is more flexible. Both dimensions are viewed as continuums that run between minimum and maximum points, allowing for multiple positions in between, which permits nuanced and differentiated definitions. For instance, "weak states" are typically found in the macroarea with low capacity and low autonomy, but those at the intersection of minimum capacity and autonomy would be defined as "failed" states, while those close to the medium points are hybrids on the verge of change. Similarly, a "strong state" positioned at the top-right end of the spectrum—that is, at the intersection of maximum capacity and maximum autonomy—is likely to be "despotic," but states situated below that point within the same quadrant can be considered just effective states. On the flip side, a capable state situated at the intersection of maximum capacity and minimum autonomy appears more like a mere "instrument" in the hands of external powers. Furthermore, many empirical cases are expected to fall somewhere in the medium positions identified by the dotted lines across the quadrants. These intermediate cases that only partially approximate the ideal types should not frustrate us—rather, they are probably the most interesting cases to watch. Overall thus, our choice is to abstain from imposing too stringent conceptual constraints onto reality. In a measurement-oriented approach, the location of a state within a quadrant—whether at the centre, edges, or fluctuating somewhere in the middle—has a significant impact on assessing its status, calculating its trajectory, and predicting future changes. And after measurements have assigned positions in the space, a significant amount of interpretation remains to be done.

This said, the next section will focus on exploring the "forgotten core" of state power, that is, the autonomy dimension.

2.2.2 State Autonomy (and Its Enemies): The Forgotten Core

The notion of state autonomy as a constituent dimension of state power encompasses two distinct subdimensions. The first, which is referred to as "hard" or technical autonomy in this book, pertains to a state's material self-sufficiency, the ability to operate without relying on foreign states, potentially stretching to a state of complete autarky. The second subdimension, termed "soft" or political autonomy, describes a state's ability to define its national interest and act on it independently or even against competing interests from both within and outside the state, potentially extending to the extreme of despotism. Chapter 3 will address the technical subdimension, which is best defined in terms of policy specifics. In this section, instead, we will focus on the political subdimension by presenting an overview of various studies that examine the degree to which states are shielded from or exposed to the influence of foreign states as well as domestic actors, primarily the military and private corporations. These studies are scattered across various disciplines; they are rarely couched in theoretical language or strive to develop metrics of state autonomy. When taken together, however, they illustrate how political autonomy, or the absence thereof, is a crucial factor in understanding and measuring state power.

2.2.2.1 Definitions: Actors, Decisions, Domains

Measuring the political autonomy of the state in a comparative perspective requires developing conceptually sound indicators that can identify (a) *in which domains* state autonomy is being upheld or challenged, and (b) *who* has the potential to challenge it, *how* and *why*, i.e., the "agents of influence," their resources and modes of operation.

In order to operationalise the first question, we distinguish two broad domains of State activity—*external* decisions and *internal* decisions—and focus on how these decisions are made rather than on their nature, content or scope.

- *The domain of external decision-making* refers to the process of determining a state's political actions in the international arena. In particular, we concentrate on the autonomy of the state in (a) electing to participate in international agreements, treaties, and organisations; (b) choosing a course of action (such as voting, vetoing, and forming coalitions) while part of multilateral forums or organisations; and (c) deciding whether to comply with international commitments to which the state has officially pledged itself, including legally binding law and norms, like treaties, or more flexible mechanisms such as regulations, guidelines, best practices, codes of conduct, or technical standards.
- *The domain of internal decision-making* pertains to domestic decision-making, specifically focusing on the autonomy of the state in (a) formulating public policies of major national relevance; (b) designing operational programmes for policy implementation; (c) selecting partners for policy implementation and programme execution, which includes the allocation of resources and incentives, e.g., public contracts, public–private partnerships, domestic-foreign cooperation arrangements, etc.

Regarding the second question, there are various types of agents of influence that have the potential to significantly challenge state autonomy. These too can be categorised as either *external* (foreign) or *internal* (domestic), and further differentiated based on whether their interests and resources are rooted within the state's institutional sphere or lay outside of it. This can be operationalised as follows:

- *External agents of influence* include foreign states and multinational corporations.[13] The importance of the former is self-evident, as the fundamental principle of Westphalian sovereignty prescribes a nation-state's ability to remain independent from peer-level institutional actors, such as other states. However, this state-centric approach has been dramatically altered by the globalisation movement of the late twentieth century, due to the emergence of gigantic multinational corporations with the resources to sway, capture, or even replace nation-states to push their own interests and agendas.
- *Internal agents of influence* consist of the military and domestic corporations, whether private or state-owned. The impact of private corporations in this regard is apparent and has been extensively studied: influential economic powers whose interests lay squarely outside the public sphere have always tried to hinder or weaken the autonomous power of the state in policymaking. In turn, the state's

[13] This is of course a simplification. A comprehensive list of agents of influence operating at this level should include at least major international institutions and organisations such as the International Monetary Fund (IMF), the World Bank, and the World Trade Organisation (WTO), as well as independent global civil society actors such as NGOs. We justify our choice to exclude these actors from our analysis not only on the basis of the benefits of simplification, but mainly by their relative lack of relevance for an analysis of power in the space sector.

ability to resist and counteract this influence is crucial for asserting its sovereignty domestically. The roles of the military and state-owned enterprises (SOEs) are more complex, given their hybrid nature. On the one hand, they are an essential component of the state, and their leaders constitute a bureaucratic-technocratic establishment closely linked to the political (civilian) elite to whom they are accountable. On the other hand, both the military and SOEs may develop their own organisational interests and priorities, which stem from their functional domains (*force* and the *economy*) and may conflict with the "general interest" as defined by the state elites who are meant to govern and direct them. In such instances, both actors may act as powerful agents of influence on the state "from within."

All of the above is summarised in the Table 2.1.

Unravelling the complex network of influence relationships between various types of state decisions and agents of influence is a challenging task. The table above may suggest that external agents tend to focus on international decisions, while internal agents concentrate on domestic ones, but reality is much more intricate. States can both exert influence and be influenced, and their sovereignty can be threatened from both external and internal sources which may affect decisions in both domains. A state may prove autonomous or even dominant vis-à-vis other states in external decisions (for example by imposing international alignments and coalitions) as well as in internal decisions (an extreme instance would be foreign sponsored "regime change"). But the same state may prove vulnerable to domestic agents of influence, such as the Armed Forces and business corporations, in the realm of domestic decisions. Additionally, while the military and businesses of a country typically do not influence foreign states directly, they can indirectly do so by inducing their governments to act on their behalf. In summary, analysing the political autonomy of a state and the factors that challenge it necessitates a multidimensional approach that renders the task highly complex.

In what follows we begin to lay the groundwork for such an endeavour. Because in this book we use a survey to measure qualitative matters such as state autonomy (or lack thereof) in the realm of space policy (see Sect. 2.3), what we need to do here is provide the rationale for the choice of the questions that we posed to our select group of experts. To achieve this, we begin with the more general level of state power—of which space power is a specific manifestation. We will show that a sufficient

Table 2.1 Types of decisions, agents of Influence and decisional domains

	External domain	Internal domain
Types of decisions	Joining treaties and organisations Acting within multilateral fora Complying with norms, rules, and procedures	Formulating major national policies Devising implementation programmes Selecting operational partners
Agents of influence	Foreign states Multinational/Transnational Corporations	National military establishment Domestic business corporations (public and private)

amount of empirical material and scholarly analyses exits that support asking those questions in order to identify and measure state autonomy as a component of state power. Subsequently, we turn to the literature on space power to examine how these findings can be adapted and applied to our specific field of analysis.

2.2.2.2 Challenging State Autonomy in External Decisions: Inducements and Coercion

We have defined the autonomy of a state in the context of external decisions as being associated with the ability to make independent choices regarding the political conduct in the international arena. This does not come without challenges, however. According to Joseph Nye (2004: 2), "there are several ways to affect the behaviour of others. You can coerce them with threats; you can induce them with payments; or you can attract and co-opt them to want what you want." International interactions typically involve a combination of these three approaches, but for analytical purposes, it is necessary to differentiate between them. Here we will disregard the third approach, which Nye termed "soft power," due to its pervasive but elusive nature. Instead, we will focus on *coercion* and *inducements*, which represent the two faces of "hard power."[14]

Accordingly, state autonomy can be measured by the degree to which a state can establish its political conduct independently of the will of actors who have the capability to use inducements or threats to challenge its decision-making authority in the external domain. Considering the resources required to raise such a challenge, these potential "agents of influence" will typically be other states, although, as we will explore, other actors may indirectly enter the fray when their crucial interests are involved.

State-to-State Interactions: The Practice of Vote Buying

During the height of the escalation leading to the first Gulf War in 1991, the United States proposed a resolution to the UN Security Council authorizing the deployment of armed forces against Iraq. To secure support or abstention votes, financial aid was promised to four developing countries that were rotating members of the Council, while a fifth, which opposed the resolution, was penalised with a sensible reduction in annual aid. Additionally, the United States attempted to purchase the votes of permanent members. It promised to persuade Kuwait and Saudi Arabia to provide the Soviet Union, which was on the verge of collapse, with hard currency "to make overdue payments to its commercial creditors." For China, which had been severely

[14] Note that to Nye (and to us) both faces of power are fundamentally coercive, as they are based on commands rather than persuasion, though they both combine sticks and carrots. In its military form, coercion works through force and threats as well as through the offer of protection, as in alliance-building. In its economic form, it may entail applying sanctions as well as offering payments, the latter ranging from aid funds to plain bribes (Nye, 2004: 8, 31).

sanctioned for the 1989 Tiananmen Square massacre of pro-democracy protesters, the United States offered to lift trade sanctions, support a remarkable World Bank loan, and resume "normal diplomatic intercourse between the two countries" (Eldar, 2008: 17).

This episode is extreme, due to the critical importance of the issue on the table and the massive economic resources mobilised by the US, either directly or through affluent client states and Washington-based international financial institutions. But buying votes is an ordinary practice in state-to-state dealings across a range of issues and organisational contexts. A significant amount of literature exists that uses vote buying to operationalise the concept of "inducements" in international power relations. For our purposes, vote buying is a useful tool to exemplify challenges posed by one state (the buyer) to the political autonomy of another (the seller) when it comes to *acting within multilateral organisations.*

A well-known instance of this practice is the "vote-for-aid" exchange taking place in the UN General Assembly (UNGA). For decades, the United States has been a major player in this regard (Bueno de Mesquita & Smith, 2007; Carter & Stone, 2015; Palmer et al., 2002). Indeed making "strategic use of foreign aid" is an old American doctrine, and the US has been regularly employing bilateral aid agreements as a political punishment, reward, or incentive to influence voting behaviour in the Assembly. Receiving countries, on the other hand, tend to either passively abide to pressure and take the bribe or to actively sell their vote to the highest bidder (Alesina & Dollar, 2000; Alesina & Weder, 2002; Woo & Chung, 2018). While the practice dates back to the Cold War era, recently China has emerged as a serious competitor. The frequent voting alignment of the "African Bloc" with China is notoriously connected to the latter's massive economic presence on the continent (Strüver, 2016).

The US and China also trade aid funds to gain influence among third world countries that are non-permanent members of the UN Security Council (UNSC). Recent research shows that China employs its most prominent state-owned enterprises to systematically buy influence on the UNSC for its foreign policy strategy (Stone et al., 2022: 18). Meanwhile, the US increases aid to temporary seat holders by more than half during their tenure, making membership more attractive, and gratitude to the US compelling. The US also indirectly channels additional funds through UN agencies it can influence, such as UNICEF, as well as through the World Bank and, to a lesser extent, the International Monetary Fund's "loans for votes" strategy (Dreher et al., 2009a, b; Kuziemko & Werker, 2006).

As Vreeland (2019: 212) summarises these findings:

> Scholarship on the Bretton Woods institutions … reveals that powerful countries use the resources of these organisations, which are intended to address macroeconomic imbalances and economic development, to pay off favored countries. … [T]he fact that IMF and World Bank money goes to countries with specific voting behavior at the UN implies that the votes are bought and sold.

A final example concerns tactics used in the World Trade Organisation (WTO), whose voting system is based on unanimity rather than majority vote. Since each country has a veto power, reaching a consensus would be impossible without employing informal exchange and coalition building techniques such as "vote trading," "logrolling," and "coercive tendering," among others. Analysts have ascertained that these deals take place in closed-door meetings where different forms of pressure are employed, from simple inducements involving collateral payments, to coercive tactics and outright threats, including withholding of preferential trade agreements or blocking the availability of loans from the IMF (Eldar, 2008: 27, 29).

To summarise, scholars have gathered ample statistical and anecdotal evidence of vote buying in international organisations and forums, and it is notable that they routinely refer to it by employing the language of bribery. In fact, if this practice took place in a domestic context, it would undoubtedly be classified as corruption, which is banned by law in most countries (Vreeland, 2019).[15]This is not the place to enter this discussion, if not to underline that stigmatizing vote buying as an illegal behaviour makes it appear "exceptional"—not only a deviation from legal norms, but from "normality" itself. Instead, as Joseph Nye puts it, "inducements" are a common method of exercising power. It may not be perceived as illegal or abnormal in many socio-cultural contexts, and as a matter of fact it is widely practised in international relations, even by states that prohibit it domestically. Be that as it may, for our purposes here it will suffice to note that when states use wealth in this way, they act as agents of influence impairing the ability of other states to independently determine their own political behaviour in international arenas.

In what follows we will see that inducements can be associated with "threats" when it comes to matters of military concern and that the two faces of hard power can sometimes work in unison. We also show that when this happens, the domestic military establishment of the state that carries out the threats enters the influence game. Essentially, this means that the same civilian elites who appear to hold sway in international relations may not have the same degree of political autonomy domestically, particularly in matters that impact the interests of their military.

Military Influence: The Case of US Opposition to the ICC

Deciding whether or not to become a member of international treaties and organisations, and whether or not to comply with the associated formal obligations are vital issues in establishing a state's autonomy. These figured prominently in the United States' effort to boycott the International Criminal Court (ICC) during the

[15] Some legal scholars suggest that formal rules should be adopted by the UN to prohibit vote buying, as it conflicts with the principles of good governance that are supposed to form the foundation of the international order (Gillespie, 2004:104). Others have emphasised, however, that the international community has not expressed concern over vote buying as both buyer and seller states have an interest in maintaining the practice as legal (Lockwood, 2013: 100). Still others have argued that vote buying has benefits, as it redistributes resources from wealthy to poor states and serves as a mechanism for "welfare maximization" by vote sellers seeking rents (Eldar, 2008: 16). As argued by Pamela Karlan (1999: 1711) voting by public officials is a public function, not solely a private right and should not be commodified. This principle applies to both domestic and international contexts: alienating decision-making authority in exchange for economic gain is equivalent to selling state sovereignty.

early 2000s by inducing other states *not to join* or *not to comply with* the international treaty known as the Rome Statute. While the United States did employ economic pressure to this effect, by warning that it would withhold development assistance from countries that did not align, its tactics also took on a menacing tone with threats of withdrawing military aid and support. Although this approach was not a direct military threat, it could still be considered an indirect one, given that many of the targeted nations relied on US military aid to protect themselves against war or insurgency. Leaving them out in the cold could result in their defeat and possibly even state collapse. This exemplifies Nye's concept of the "first face of power."

The International Criminal Court was established by the Rome Statute in 1998 to investigate and prosecute individuals accused of war crimes and other crimes against humanity. Former US President Clinton, who had previously supported the Court, signed the Treaty on his last day in office on 31 December 2000. However, he expressed "deep reservations," succumbing to mounting suspicions in the US that the Court could become politicised and put in jeopardy American troops deployed abroad as well as the country's national sovereignty. The Republican-controlled Congress that followed did not ratify the treaty, and in May 2002, President G. W. Bush "unsigned" it with a letter addressed to UN General Secretary Kofi Annan.

Despite US opposition, the Rome Statute came into force in July 2002. Shortly thereafter, the Congress passed the "American Servicemembers Protection Act," which prohibited the US government from cooperating with the Court and providing military aid or assistance to countries that ratified the Statute. An exception was made for countries that, despite becoming members of the ICC, agreed to enter into a "Bilateral Immunity Agreement" (BIA) with the US, pledging not to surrender American citizens to the Court. In essence, BIAs aimed to nullify the Rome Statute by committing signatories not to comply with its provisions when US armed forces were involved (Kelley, 2007: 575). This posed a direct threat to the political autonomy of other states, indeed to their very sovereignty. As the Executive Director of Human Rights Watch denounced in an open letter to the US Secretary of State Gen. Colin Powell:

> Many states, including close US allies, signed [BIAs] only after having been threatened and coerced. Officials from a number of governments have stated publicly that they believe the agreements violate their international treaty obligations, their domestic laws and in some cases even their constitutions. Several states have signed [these] agreements only in the face of what their diplomats have labelled 'unbearable' pressure, including threats to cut not only military aid, but humanitarian aid and economic assistance as well. (Roth, 2003)

Not surprisingly, countries that were wealthier and less reliant on US military aid proved more likely to decline signing a BIA. None of the EU member states, Canada, Australia, or Japan signed a BIA. Facing isolation from its own allies, the US attempted to lessen the damage by granting a "national interest waiver" to NATO members and other significant non-NATO partners in the War on Terror. The majority of those who obtained the exemption were developed, democratic nations that belonged to the OECD (Nooruddin & Payton, 2010). In short, when challenged by a powerful foreign entity, certain states were able to protect their political autonomy, while others were not and succumbed to pressures.[16]

That said, the aggressive US stance on the ICC issue reflected, at least in part, widespread military concerns. These were especially influential in a post-9/11 world where the US was waging a broad and controversial War on Terror in the Middle East and the Secretary of State was (unusually enough) a four-star general and former Chairman of the Joint Chiefs of Staff. As reported in a military journal:

> U.S. military opposition to an International Criminal Court was both predictable and influential. … The Pentagon had powerful friends in the U.S. Senate where any negotiated treaty must be submitted for advice and consent. … The U.S. Senate Foreign Relations Committee (specifically the Subcommittee on International Operations) considered the ICC to be "the most dangerous threat to national sovereignty since the League of Nations" (Hull, 2012: 61).

The issue of military-civil relations in the United States has been extensively studied and is documented in the literature.[17] In essence, although the military is considered to be a "part of the state" and primarily dedicated to national interests, its perspectives may not always align with those of civilian elites. Additionally, like any organisation, the military has its own vested interests to safeguard and internal cohesion to nurture. When these are at risk or perceived to be so, the military has an incentive to act as an independent agent of influence. The effectiveness of this dynamic largely hinges on the degree to which the civilian elite in power is politically and culturally attuned to military positions and interests. Consequently, the seemingly symbiotic relationship between military and civil elites fluctuates between collaborative and confrontational modes. The case of the ICC is illustrative in this regard. As we have seen, President Clinton did sign the Rome Statute, although with reservations, while the Bush administration adopted an unequivocally hostile stance. Going further, President Obama cooperated with the Court by offering rewards for the arrest of fugitives indicted in Africa, while President Trump reverted to hostility and imposed sanctions on ICC personnel investigating the torture of terrorism detainees during the Bush era. Notably, in March 2023, when the Biden administration hinted at collaborating with the ICC to prosecute Russian President Putin for war crimes in Ukraine, the Pentagon came close to openly opposing the idea due to concerns about setting a precedent for prosecuting Americans. Bureaucratic infighting emerged in Washington, with the Department of Defense and military leaders opposing assistance to the court, while "the rest of the administration, including intelligence agencies and the State and Justice Departments," favoured it (Savage, 2023).

---

[16] It is worth noting that there is an interesting similarity between this case and the instances of vote buying previously examined. In fact, US policymakers justify both military and development aid to foreign governments on the premise that doing so "will increase US influence over the recipients' foreign or domestic policies" (Sullivan et al., 2011: 277). In our terminology, in both situations, the core of the latter's stateness is brought into question.

[17] The literature is abundant but, significantly enough, its two unshakable pillars remain Samuel P. Huntington's *The Soldier and the State* (1957) and Morris Janowitz's *The Professional Soldier* (1960).

Ultimately, the ICC case reveals an intricate scenario with multiple players involved in a reciprocal game of influence. Firstly, the majority of the international community, that supported the ICC and was seen by the US as a threat to its sovereignty. Secondly, the US government, which challenged the sovereignty of other nations as well as their domestic laws and constitutions. Lastly, the US military, which considered the ICC as a menace to their prestige and the safety of their troops, and acted as a domestic agent of influence vis-à-vis their own government, indirectly affecting the political autonomy of foreign states as well. This confirms our assumption that the role of the military must be taken into account when assessing state autonomy.

We will now explore another domestic actor that has the ability to exert a combination of direct and indirect influence on both external and internal decisions—business corporations.

Business Influence: The Case of Japan's Whaling Triangle

The International Whaling Commission (IWC) was founded by a group of states in 1946 as a voluntary international environmental organisation. Its mission was to ensure the proper conservation of whale stocks and facilitate the orderly development of the whaling industry. In 1982, after around 40 years of operation, the IWC implemented a moratorium on commercial whaling for all whale species and populations, which remains in effect today. Japan was one of the strongest opponents to the ban and attempted for decades to overturn it by using its Official Development Aid (ODA) programme to purchase the votes of weaker IWC members. A group of Eastern Caribbean states, with their vulnerable economies, were a primary target for Japan's vote-buying strategy. A study has shown that these states consistently sided with Japan in IWC votes due to the disbursement of ODA funds (Strand & Tuman, 2012).

This well-known, much-studied case prompts broader considerations on the interaction between foreign and domestic agents of influence and the interdependence of internal and external decision-making domains. While differentiating between these domains can be a useful analytical tool, it is important to recognise that they may be more intrinsically intertwined than might be expected.

The situation involving vote-buying in the IWC, in fact, demonstrates that while Japan appeared to be a dominant foreign influencer in relation to smaller and weaker nations, the Japanese state itself was vulnerable to a potent lobby with strong connections to influential government officials, known as "The Whaling Triangle." This group was dominated by vested interests led by the Japan Fisheries Association, which "not only influence, but also direct the nation's whaling policy" (Kagawa-Fox, 2009: 401; Morikawa, 2009). The phrase alludes to the "Iron Triangle," a traditional Japanese institution which "comprises a close relationship between government bureaucrats, politicians, and big business held together by common interests" (Kagawa-Fox, 2009: 403). This connection is further cemented by the institutionalisation of *amakudari* (meaning "descent from heaven"), a Japanese version of the revolving door phenomenon whereby senior bureaucrats who retire are parachuted to prominent positions in public institutions and private firms. These people "can be as powerful as politicians in influencing a ministry ... [for] they have access 'to currently serving bureaucrats who were previously their juniors in ministries and who remain susceptible to their former superior's influence and persuasion'" (Mulgan, 2000, quoted in Kagawa-Fox, 2009: 404). The intricate networking established through informal personal connections enables the business interests that control the Triangle to shape national policies from behind the scenes.

Interestingly, when applying this model to examine Japan's actions in the IWC, two researchers employed the language of state autonomy, specifically the absence of it:

> Parts of the Japanese state enjoyed autonomy, but the Ministry of Agriculture, Forestry and Fisheries (MAFF) and the Ministry of Foreign Affairs (MOFA) have been vulnerable to pressure from cohesive groups in the fisheries sector. Detailed evidence [suggests] that close contact between current bureaucrats in MAFF and MOFA and retired colleagues who work in the sector have facilitated consensus and made government policy more responsive to sectoral demands. Japan's recession also weakened the autonomy of the state [vis-a-vis influential business groups, including those in the "lucrative whaling market"] (Strand & Tuman, 2012: 410, fn. 3).

The interplay between these domestic power dynamics and their international implications, such as the use of development aid to secure votes in the IWC, illustrates the complexity surrounding the concept of state autonomy when the role of business corporations comes to the fore.

Concluding Remarks

A few observations can conclude this section and introduce the next one. The first regards the *connection* between domestic and foreign policy arenas, whereby agents of influence in one country indirectly affect the behaviour of foreign countries by shaping the policy agenda of their own government. For instance, the military has been identified as a driving force behind the United States' decision to threathen foreign states into violating their commitments to the Rome Statute. Similarly, the Fisheries Association was the group behind Japan's vote-buying strategy in the International Whaling Commission.

A second observation concerns the *distinction* between domestic and foreign domains, emphasizing that a country's dominance on the global stage does not necessarily translate into a high level of state autonomy at home. Thus, global and regional powers such as the United States and Japan, that are powerful enough internationally to buy up or coerce other states, domestically may be vulnerable to forces from within the state, such as bureaucratic or military pressures, and/or from outside, such as big-business pressures.

Thirdly, state autonomy and vulnerability are relative rather than absolute notions, and vary across different decision-making arenas. In some policy domains, certain "parts of a state" may possess a high degree of autonomy, while other "parts" involved in other policy decisions may be captured by powerful exogenous interests. This occurs because different policy arenas activate different actors with diverse interests, resources, and leverage on the government.

Last but not least, we should consider the importance of *the nexus between wealth and power*, which can influence a country's domestic and foreign policies alike. This is evident in international vote buying, where wealthy nations trade development aid to induce poorer countries to sell their sovereignty *à la carte*. On the domestic level, this connection has emerged in the example of Japan, where powerful private interests are able to capture the state, or "parts" of it, effectively dominating public decision making in a specific policy area. We now turn to this last line of reasoning by reviewing the major literature on the domestic challenges to state autonomy posed by business corporations.

2.2.2.3 Challenging State Autonomy in Internal Decisions: The Wealth-Power Nexus

This section turns to the national context and examines the extent to which state elites can make strategic policy decisions independently of or against the interests of domestic business corporations. Stated the other way around, it explores the extent to which firms can influence or even capture public decision-making to advance their interests over the public interest, and how states respond. Our attention therefore focuses on state-business relationships in the public policy decisionmaking process: from the formulation of strategic objectives to the adoption of regulatory decisions, including implementation programmes, to the selection of industrial partners and the allocation of public resources. In short, we try to identify the differences and similarities in state autonomy and corporate influence among the countries in our sample.

Some caveats are in order. First, we start from the assumption that "what is good for businesses is not (or not necessarily) good for the country," and that what is good for a country is still up to the public authorities to decide. We therefore view business corporations as agents of influence potentially challenging state autonomy. Second, elaborating on Joseph Nye's definition of "inducements" as a fundamental resource of hard power, we assume that wealth is the main political asset available to economic actors. So our guiding research questions are: what role does wealth play when it comes to determining strategic public policies? Where does a given state stand on the weak-strong state continuum when it comes to the nexus between wealth and power? The third caveat is that the countries chosen for this examination are part of the sample used in this book to assess state power in the field of space. The sample comprises eleven nations chosen for being among the most active in space over the previous decade (see Chap. 4). These include (a) most of the "Global West," which comprises the United States and its allies in Europe (taken as a whole), Israel, and the Pacific region (Australia, Japan, South Korea); and (b) the most powerful emerging economies of the BRICS (Russia, China, India, and Brazil). In what follows we will focus on three countries that may be considered relatively representative of these two groups: the United States and two BRICS countries, India and Russia—the former being a somewhat imperfect democracy, and the latter an autocracy, itself far from perfect.

Wealth and Power in the BRICS

BRICS nations share a common history of inward-looking, state-controlled industrial development, and have adopted different models of state capitalism that vary depending on whether they are located in communist (China), post-communist (Russia), or "imperfect democratic" environments (India, Brazil, South Africa). However, globalisation policies, such as liberalisation, privatisation, and deregulation, have eroded the state-centric nature of state capitalism, favouring the interests of both national and multinational corporations. In situations where impersonal norms and the rule of law are weak, and clientelism and corruption are widespread, economic growth and political stability can be more effectively achieved through interconnected patronage networks rather than through the legal-rational institutions of a Weberian state. Not surprisingly thus, the literature frequently refers to these political-economic systems as "crony capitalism" rather than state capitalism. The notion of crony capitalism denotes an economic setup where politically connected businesses receive preferential treatment from the government, including protection of their assets, in return for political leaders sharing in the rents generated by these assets. The resulting interaction between business and the state fluctuates between state capture and more cooperative relationships, blurring the distinction between public and private spheres and revealing that the state may be more vulnerable to external pressures than it might seem. This takes different forms in different countries.

India

After the 1997 global financial crisis, "crony capitalism" has been identified as a major Asian malaise (Haber, 2002: xi) and the relevant literature sees it as a peculiar feature of corporate-state relations in India. Indian scholars highlight that their country's version of cronyism is oligarchic and familistic (Varma et al., 2016) deeply embedded in Indian society and reveals a "misalignment" between formal norms and the underlying value system, where the public–private divide is not recognised (Billing & Farro, 2016). Furthermore, India's cronyism has a remarkably long history and has demonstrated considerable ability to adapt to changing political and economic conditions. According to a group of Indian scholars, its origins date back to the pre-colonial era, and the core of the model remains "largely intact even today" (Khatri & Ojha, 2016: 63).

As a matter of fact, the state-run economy of the first post-independence era (1947–1991) has been described as a form of crony *socialism* (Mazumdar, 2008). It was rooted in the extreme regulatory power of the state, leading big business families to focus on "managing the government" in order to secure quasi-monopolistic rents

(Varma et al., 2016: 162). Crony *capitalism* proper emerged in the 1990s from the shift towards a more liberalised and globally interconnected economy. The privatisation and relaxation of state regulations played a critical role in this transition, generating "numerous opportunities of conferring benefits on businesses having a privileged relationship with decision-makers" (Mazumdar, 2008: 13). This effectively resulted in a "legalized carve-out" of the state (Khatri & Ojha, 2016: 62).

In this cultural and institutional setting, corporate influence over state officials operates through patronage networks and behind-the-scenes interactions, rather than through open collective action or policy advocacy (Kumar, 2016: 105, 111). This creates a level of interdependence between wealth and power that borders on a symbiotic relationship, a phenomenon that is especially evident when looking at traditional multinational tycoons like the Hindujas and Mittals, as well as at emerging, aggressive middle-class entrepreneurs in the field of information technology. The impact of this dynamic extends beyond domestic politics to shaping foreign policy, particularly foreign economic policy. On the one hand, globally oriented business elites act as intermediaries for Indian officials abroad, as they know how to "open doors in the corridors of power" in international capitals and "lubricate the system" of informal foreign relations by financing foreign politicians, foundations, and interest groups (Baru, 2009: 273–74). On the other hand, this proximity to businesses also poses a potential threat to the integrity of Indian state institutions, as civil servants in the Foreign Ministry may be tempted to demand or accept rewards for the introduction services provided to the private sector, which can range from bribes to post-retirement appointments in private firms (Rana, 2017).

Russia

In post-Soviet Russia too, the nexus between business and the state is frequently characterised as crony capitalism, though it varies from India's version in significant ways. Despite sharing a profound entrenchment in informal practices and institutions and a remarkable adaptability to different contexts,[18] Russia's crony dynamics are notably more politically confrontational than India's. This becomes evident when examining the two distinct phases of Russia's post-communist transformation.

The first phase (*destatisation*) began with the "crony privatisation" of the 1990s, which resulted in the wholesale theft of public resources. It is at this juncture that

[18] Cronyism in Russia has a long history, but it was institutionalised during the Soviet era (1917–91) when it was rooted in the one-party state and the planned economy. In Moscow, behind the facade of a totalitarian monolith, informal bureaucratic and regional interest groups coalesced to influence central decision-making This system was famously labelled "institutional pluralism" or, in the Chinese variant, "fragmented authoritarianism" (Hough, 1977; Lieberthal & Oksenberg, 1988). In the periphery, "cliques of mutual protection" formed to manipulate policy implementation in their fiefdoms. These groupings relied on vertical client bonds controlled by the Communist Party, leading to impressive growth and the appearance of a strong state. The inherent weakness of this arrangement only became apparent in the 1970s and 1980s when the whole system spun out of control and ultimately led to the collapse of the state (Hough, 1997).

former Soviet industrial executives and communist party officials joined forces with a younger generation of ambitious entrepreneurs, giving rise to groups of "oligarchs" and regional "barons" who effectively took over substantial portions of the state (Hoffman, 2002; Stiglitz, 2002). Once they had amassed immense fortunes and gained control of strategic industries and the media, they used these resources to shape policy making, regulatory, and legal environments to their advantage, generating concentrated rents at the expense of the rest of the economy (Hellmann & Kaufmann, 2000: 1). The political autonomy of the Russian state reached the nadir when some oligarchs attempted to enter the political arena directly.[19] According to Mikhail Afanas'ev (2004: 4; 2009), at the end of the 1990s Russia suffered from "an unbearable lightness of the state." In a series of insightful essays Vladimir Shlapentokh (1996, 2003, 2007) conceptualised Yeltsin's Russia as a "feudal society" plagued by widespread privatisation of public power and a neo-patrimonial system of rule based on personal ties of mutual dependence, or "suzerain-vassal relations." Eventually, such situation was bound to provoke an intense reaction from the state's elites. Drawing on Shlapentokh's historical analogy, a polity aiming to move beyond feudal fragmentation found its options limited by its most pressing functional need, namely, "monarchical absolutism." This is indeed where Vladimir Putin entered the picture (Cappelli, 2010).

The second phase of the Russian transition (*authoritarian state building*) is strictly connected to the traditional Russian rhetoric of *gosudarstvennost'* (stateness). This entailed repressing a handful of insubordinate oligarchs and co-opting a new elite stratum to manage the state and economy. Known as *siloviki* ("the people of force") they came from the military, security, and intelligence sectors and were initially portrayed as selfless officials dedicated to the national interest. However, they would prove to be just as rapacious as their predecessors, although more politically compliant (Orttung, 2004: 59). At the highest level of government, this elite group now dominates the country's most profitable state-owned enterprises. At the lower levels, thousands of *siloviki* have become private businessmen while maintaining contact with their former colleagues in service. These connections are "based on mutual understanding and assistance," forming "fraternities" that lead back to the Kremlin, where members use their influence to shape policy in their favour (Easter, 2008: 210; Kryshtanovskaya & White, 2003: 302). Around this militocratic core, a politically connected state-business class emerged, made of former Soviet-era managers, young technocrats and financial executives, heads of state-owned enterprises, and leaders of the military-industrial complex. These people, collectively—and often superficially—dubbed the "new oligarchs," were forced to accept a more organic *symbiosis* between wealth and power in which the former was guaranteed by the latter based on new rules of the game, first and foremost political loyalty. The system now operates within an integrated framework, with control levers in the

[19] The most notorious among them was Yukos boss Mikhail Khodorkovsky, who was known to spend lavishly on lobbying and was reported to have over 100 Duma legislators on his payroll. In 2003 Khodorkovsky led the assault at the heart of the state by hinting at the prospect of running for president after Putin's initial term expired.

Kremlin and day-to-day functioning ensured by Putin's party United Russia. Also known as the "party of power" this is a conglomerate of Tammany Hall-style political machines which combine getting out the vote with "a mechanism for extracting rents and distributing patronage [...] financed with resources contributed by interests pursuing benefits from the State" (Remington, 2008: 960).

The foregoing, however, does not imply a complete convergence of interests between political and business elites. This is apparent even at the highest echelons, where top executives of large state-owned enterprises de facto participate in strategic policymaking while still retaining "their own corporate identities and interests." Although the political leadership regards them as instruments of their geopolitical strategy, major corporations like Gazprom and Rosneft do not always behave as obedient arms of the state and may be able to pursue their preferences even when they do not align with those of the Russian president (Orttung, 2006; Duncan, 2007:15).

In conclusion, while the notion of state capture by defiant oligarchs may no longer reflect the prevailing power dynamics in Putin's Russia, the new state-business symbiosis masks an intricate tangle of *collusion* and *collision.*

Wealth and Power in the Global West

Compared to Russia and India, most countries of the Global West enjoy a higher degree of "Weberian" stateness. The modern state is, after all, a peculiar Western formation, as is liberal democracy—hence the saying "no state, no democracy": functioning democracies need strong, effective states, both capable and autonomous (Linz & Stepan, 1996). But Western states too (or "parts" of them, or specific public policy areas) are often said to be "captured" by special interests. Furthermore, globalisation has substantially reduced the state's control over powerful domestic and multinational corporations, most of which are headquartered in Western countries. As a matter of fact, the very debate about state autonomy or lack thereof grew out of analyses of the Western "capitalist state." And neoliberalism is, again, a Western creation as well. It should be unsurprising, therefore, that the literature on state-business relations in the West abounds with terminology that we have already encountered thus far, including "state capture," "predatory corporate behaviour," "oligarchic domination" or "collusive relations" between public and private. This is particularly true for the United States, where the wealth-power nexus has been extensively studied and debated.

The United States

Since the end of World War II, American scholars have been concerned that economic interests could wield too much influence on the political system and threaten American democracy. In the 1950s, C. Wright Mills popularised the idea of a close-knit "power elite" made up of businessmen, military leaders, and state officials who effectively ruled America. These were held together by a combination of social status, wealth, and institutional position, which was un turn reinforced by common backgrounds, shared interests, and informal social interactions. The elite theory cast more than a doubt upon the American democratic creed, founded as it is on the principles of free elections and market competition (Wright Mills, 1956).

In fact, scores of mainstream scholars rushed to counter the elitists' argument. Proponents of rational choice theory contended that American politics does reflect the will of the citizenry, as politicians seeking votes tend invariably to align with the preferences of the median voter (Downs, 1957). Supporters of the pluralist theory maintained that the public interest is safeguarded in any case by the existence of multiple competing interest groups, with no one group consistently prevailing over the others. This latter perspective became the dominant paradigm of US politics under various labels, including "elite pluralism," "democratic elitism," and "polyarchy" (Dahl, 1961, 1971).

To be true, alternative perspectives on this matter have always existed in serious scholarship. Schattschneider (1960: 124) advocated that business power be "matched with governmental power." Charles Lindblom (1977: 356), himself a founder of the pluralist school, underlined that the "disproportionate power" and the "privileged position" of large corporations made them "unfit" for American democratic theory and vision. Meanwhile, a seminal strand of literature emerged which focused on the regulatory stage of policymaking, where bureaucratic agencies play a significant role in writing the rules in their final form and monitoring their enforcement. In the early 1970s, economist George Stigler (1971: 4, 18) proposed a theory of "regulatory capture," which aimed to reveal how "an industry [...] is able to use the state for its purposes..." He found that the practice was "pervasive." This approach generated a wealth of research that gathered extensive data on agency capture by private interests, collusion between the state and businesses, conflict of interests, and even "tolerated corruption." In the late 1980s, a model of "predation through regulation" was proposed, showing how the advantages obtained by firms capable to capture the regulatory process harmed both competitors and consumers, leading to market distortions and damage to the public interest (Bartel & Thomas, 1987: 240, 259).[20]This critical debate remained somewhat below the radar during the heyday of neoliberal globalisation, but it regained momentum in the early 2000s, when regulatory capture came to be perceived as a major culprit in the global financial crisis of 2007–2009 (dal Bó, 2006; Capenter & Moss, 2014; Yackee & Yackee, 2006).

The literature agrees that the primary cause of the "collusive relationships" between business and the state is the pervasive influence of lobbying—"a legal activity aiming at ... procuring private benefits" from the government. The definition emphasises the differentiation from corruption, which implies illicit transactions and is therefore viewed as an exceptional aspect, rather than ordinary, systemic one, of

[20] Interestingly, studies on regulatory capture resulted in two opposing viewpoints. One strand, represented by liberal-reformists, exposed capture by private interests and called for a regulatory state to defend the public interest. The other strand, represented by conservative-neoliberals, viewed regulation as inherently flawed and advocated for reducing the size of the state as much as possible. The latter view became dominant during the Thatcher-Reagan years and the globalisation of the 1990s.

American politics (Igan & Lambert, 2019: 5, 12). But an ambiguous relationship between wealth and power is central to lobbying.

Numerous accounts exist detailing how lobbying operates in practice. At the highest level, lobbying money can purchase direct access to the government, including participation in drafting and amending legislation by providing technical assistance through advisory groups, commissions, and task forces. At the lowest level, it pays for activities aimed at shaping public opinion, carried out through media relations departments and public relations firms. In between lies an informal middle ground, where lobbyists use a range of capture techniques to "maintaining contacts" and establish "shared perspectives" with key state officials. These include various forms of socialisation, from occasional unofficial gatherings to the establishment of stable networks of friendship and mutual assistance, and extend to offering gifts, contributions, and "revolving door" employment. As seen before in relation to Japan and India, the latter is widely practised across the globe and forms a most potent incentive for regulators to establish an "industry-friendly" track record (Igan & Lambert, 2019: 10, 12). These dynamics of influence give rise to various forms of "cliques," colluding groups, and crony relationships that, when viewed through the normative lens of *conflict* of interest, should be regarded as harmful to the public interest, as they undermine state officials' "willingness to pursue the objectives of the government *against* those of the firm." However, once corporate and government interests become informally aligned, collusion or even corruption may be perceived as desirable, and an "optimal degree of tolerance" may be applied (Che, 1995: 379).

It is worth noting that the same concepts discussed so far also appear frequently, sometimes verbatim, in the case studies of other countries previously reviewed: the difference appears to be a matter of degree rather than kind. And a comparable similarity can be noted when taking a historical perspective on the matter. Indeed, according to Igan and Lambert (2019: 20) "The idea that powerful organisations with private interests may capture the government in order to pursue their private benefits is certainly not new. Similar ideas go back at least to Montesquieu and, later, to Marx."

The search for comparative concepts capable of travelling through time and space has led some authors to rediscover the classic Aristotelian notion of oligarchy. They argue that the American political system—like many others, today as in the past—can be analysed in terms of "the political power of the wealthy few" (Winters, 2011; Winters & Page, 2009: 732). This theory's key advantage is in its emphasis on how wealth can translate into political influence and power, while its drawback lies in its narrow scope, which centres on a particular group of individuals who only mobilise when their interests are directly impacted. In fact, even proponents of this approach recognise that the American government is besieged by a *greater number* of groups representing *broader* business constituencies pressing for public spending in *virtually all* policy areas where they stand to gain. Crucially, these profit and rent-seeking motivations require a state that is willing and capable, though *not* autonomous, in planning and implementing public spending policies. This requirement may even clash with the "ideological motives" of the extremely affluent, who tend to favour "lower levels of government spending on practically everything" (Gilens & Page,

2014: 571). All this suggests that reality would be better explained by "mixed theories" that extend beyond a limited emphasis on oligarchs and take into account the ability of wealth as a whole to capture the state for private benefit.

This was indeed the main concern of the Task Force on Inequality and American Democracy convened in 2004 by the American Political Science Association (APSA). The Task Force's final report identified wealth holders as people belonging to the 12 per cent of the "wealthiest American households" that is, those with incomes exceeding $100,000 in 2000. These brackets comprise about 30 million people including "privileged professionals, managers and business owners" among others (APSA, 2004: 2, 7)—a minority indeed, but still a far cry from an oligarchy. The report reviewed studies on the ability of "moneyed interests" to gain privileged access to politicians and bureaucrats, confirmed that the excessive political influence of "special interests" erodes citizens' trust in government, and underlined the tendency of the government "to tilt toward those who already have wealth and power" (APSA, 2004: 5, 12). The report also exposed the "dominance of the advantaged" who can "speak loudly and clearly to government" while the less privileged were left to "speak with a whisper that is lost on the ears of inattentive government officials" (APSA, 2004: 1, 9). And it concluded unequivocally that these "disturbing trends ... undermine the promise of American democracy" (APSA, 2004: 20).

In conclusion, a significant amount of scholarly research indicates that, both in the United States and globally, analysing the political autonomy of the state demands considering the connection between wealth and political power. The measure of this connection is the influence of business on the state. Let us now summarise the findings derived from the literature putting them in comparative perspective.

Comparisons

Scholarly use of similar terms to conceptualise the relationship between wealth and power in different countries provides insightful perspectives on the two dimensions of state power that we are concerned with, namely capacity and autonomy.

Regarding India, Vishal Gupta differentiates between corruption and cronyism. The former refers to the act of bribing public officials "to access or expedite public services," while the latter "occurs when *decisions* to allocate public resources are distorted by money, power, access, connections, or some combination of the foregoing." Thus, corruption perturbs the state's capacity of implementation and enforcement of existing legislations, while cronyism affects the core of the state's political autonomy, i.e., the making of "legislative, executive and regulatory actions" (Gupta, 2016: 177).

As for Russia, the team of IMF and World Bank analysts cited above made a valuable distinction between 'petty corruption' and *state capture*. While the former focuses on officials responsible for enforcing existing laws, capture targets those who have formal decision-making power. This involves buying votes on important legislation, "managing" government officials to enact favourable regulations or decrees, and inducing judges to sway court pronouncements. Again, captor firms disrupt the political autonomy of the state, as they "encode advantages for themselves into the basic legal and regulatory structure of the economy" (Hellman & Kaufmann, 2001: 2).

Finally, a similar distinction is made for the United States, though without reference to corruption; instead, it focuses on capture as the outcome of successful lobbying efforts. Noting that lobbying involves "one-to-one interaction between industry and both legislative and executive branches of the government," Igan and Lambert (2019: 5) distinguish between two consequences of lobbying: *regulatory capture*, which targets "agencies establishing and enforcing the final rules by which [the industry] needs to abide," and *legislative capture*, which aims at "the legislature, whose actions form the basis for these rules." So not only administrators but "elected representatives are also motivated by pursuing private interests of the regulated industry instead of the public interest" (Igan & Lambert, 2019: 5).

In short, the crucial distinction between capacity and autonomy, which is often obscured in theoretical discussions about "the State" or in efforts to measure state power empirically, seems to be clearly understood by lobbyists and business agents who work on the ground as well as by students of business-state relations.

It is also apparent, however, that different authors interpret the same concepts differently and work with different parameters. For instance, the term "corruption" is frequently used in discussions of India and Russia, but not in the United States. The concept of "capture" is also used in different ways. In India, it is implicit in the description of crony capitalism as a form of collusion between money and power. In the US, it refers to the influence of strong lobbying efforts. In Russia, it accurately reflected the situation of complete state control by the oligarchs under Yeltsin, while it does not adequately elucidate the nature of business-state relations that characterises Putin's regime.

This tendency to depict specific situations rather than interpret longer-term processes is common to almost all the reviewed analyses. Specifically, the concept of state capture is potent and evocative, but also presents some limitations. It implies a clear-cut dichotomy between the captor (the firm) and the captured (the state), as well as a fixed hierarchy of dominance between the two. While this may be the case in particular historical periods, or specific sectors and policy domains, it is too extreme and one-sided to effectively represent the entirety of power dynamics in more nuanced situations. Moreover, it is inadequate to explain changes and shifts in the wealth/power balance over time.

The Wealth and Power Symbiosis

To correct these shortcomings and overcome conceptual differences, we propose to turn to the concept of "symbiosis." This term too emerged here and there in most studies reviewed so far. The conventional definition of symbiosis refers to "the living together of *unlike* organisms;" it does not necessarily entail cooperation and may actually involve competition and even antagonism. In fact, biologists distinguish (among several other forms) between mutualistic and competitive symbiosis.[21] The latter is particularly promising for our purposes as it allows to differentiate between situations where one or the other organism predominates.

By way of analogy, in fact, we can compare the relationship between wealth and power, or business and the state, to a symbiotic relation that ranges from collusion (mutualism) to collision (competition). When mutualism prevails, neither party predominates and both benefit from the collaborative relationship. Conversely, in cases of competitive symbiosis, either party may strive to attain a dominant position—and which of the two prevails, in what manner, and in which policy domains bear significant implications that can only be assessed through empirical comparative analysis.

To exemplify, in India we see at work a case of *mutualistic symbiosis*—an opaque process of informal interactions among interdependent patronage networks, where state officials and well-connected business elites exchange favours and intermediation services. India also serves as a prime example of what Haber refers to as the "implicit contract" of crony capitalism, where private investments and growth can still occur despite a relatively weak state and a fragile rule of law. This happens because privileged businessmen have their property rights credibly guaranteed by a symbiotic relationship where "members of the government itself, or at least members of their families ... share in the rents generated by the asset holders." According to some authors, this setup fits particularly well the political strategy of current Prime Minister Narendra Modi, who seems inclined to "pragmatically legitimise" crony capitalism in the attempt to advance his "economic growth first" agenda (Khatri & Ojha, 2016: 77–78).

By contrast in Russia, after violently repressing the first wave of insurgent oligarchs, President Putin has tried to mould a new business class compelling it to accept a symbiotic relationship with the State. This arrangement guarantees personal enrichment and privileged access to the government in exchange for political subordination. It represents an authoritarian version of the aforementioned implicit contract, reflecting Putin's "political stability first" agenda. The arrangement aims to neutralise the risk of business-state collision through strong incentives for collusion. These include "sticks," like the threat of violence and imprisonment for non-compliance, as well as "carrots," such as the guarantee of predictability and protection within a centralised structure of rent distribution. The outcome is *a form of competitive symbiosis where political elites tendentially predominate* by controlling the patronage networks of the president's party. This notwithstanding, the ability of business elites to wield political influence and promote their policy preferences is not eliminated, for the "party of power" functions as a two-way channel of influence

---

[21] This definition was first crafted by Heinrich Anton de Bary in 1879. Although proposals to redefine the concept of symbiosis by equating it with mutualism "have led to confusion ... unfortunately, in many academic circles, the terms symbiosis, mutualism, and cooperation have similar meanings and are often used interchangeably" (Paracer & Ahmadjian, 2000: 6). Besides mutualism and competition, scientists have proposed various other forms of symbiotic associations including commensalism, amensalism, parasitism, and predation.

and is "more a target of intensive lobbying than a source of unified and consistent policy direction for the country" (Remington, 2008: 959–60).

As for the United States, similar to other Global West nations there should be no requirement for an implicit contract between the State and society for, in theory at least, the system operates under an "explicit contract" based on the rule of law and emphasizing a clear normative differentiation between public and private. In practice, however, informal extra-contractual negotiations are needed to make the system work and these are primarily driven by lobbying. Furthermore—and crucially—these efforts may go beyond regulatory and legislative capture, and also produce "intellectual capture." By this is meant that, as a result of the constant interaction between regulators and regulated, their respective mindsets end up resembling each other. In this way the industry views become *rationalised* and *institutionalised* as the regulator's view and, ultimately, "the society as a whole may start believing that what is good for the regulated industry is good for all" (Igan & Lambert, 2019: 5). Clearly this passage paraphrases General Motors Chairman Charles Wilson's famous statement: "What is good for General Motors is good for the country." This highlights the issue of *convergence* rather than *conflict* of interests between business and the state. Once such convergence becomes common sense, the language of state capture turns out to be inadequate to grasp what has actually evolved into a symbiotic relationship. The United States in fact exemplifies *a form of competitive symbiosis where business elites tendentially prevail*. The arrangement operates "by means anathema to [the] official, legal, and procedural objectivity" postulated by the Weberian ideal type, leading to the "fusion (and confusion) of state and private power" (Wedel, 2009: 39; 2014). While this setup may support significant growth, political stability, and high levels of state capacity in critical areas, it does so at the expense of the state's political autonomy. Accordingly, the United States can be described as a powerful government that nevertheless exhibits a comparatively low level of stateness.

Table 2.2 summarises the conceptualisations of the wealth/power nexus treated thus far.

Concluding Remarks

Having outlined the conceptual framework for analysing the autonomy dimension of stateness, the next step is to find a method to measure it empirically and compare it with the capacity dimension by charting them on a matrix akin to Figure 2.4 above. In the following pages, we will proceed in this direction, albeit with a significant caveat.

Our framework will not be employed for a comprehensive assessment of "state power" in a general sense; instead, we will specifically concentrate on the domain of space power and policy. There are three primary reasons for this approach. Firstly, the existing state of research and the availability of empirical indicators caution against committing the "holistic fallacy" of pursuing a universal measure. This task

Table 2.2 Conceptualisations of the wealth/power nexus

	DIMENSIONS OF STATENESS		FORMS OF WEALTH/ POWER SYMBIOSIS
	State capacity (implementation and enforcement)	**State autonomy** (decisionmaking)	
India	Administrative corruption	Cronyism	**Mutualistic symbiosis** Interdependence of political and business elites (growth-oriented 'implicit contract')
Russia	Petty corruption	State capture	**Competitive symbiosis (I)** Relative prevalence of political elites (stability-oriented 'implicit contract')
USA	Regulatory capture	Legislative capture	**Competitive symbiosis (II)** Relative prevalence of business elites (lobby-induced 'intellectual capture')

appears to be, at best, premature.[22] Secondly, the nature of state power itself resists facile generalisations, given its inherent "unevenness across policy areas" (Skocpol, 1985: 17). Therefore, only after testing the framework in major policy domains will we be able to apply it at a broader level. Lastly, and most significantly for our purpose, the domain of space power aligns remarkably well with this inductive approach, as it can be regarded as a peculiar manifestation of state power. On one hand, it is intricately linked with the realms of defence and national security, naturally falling within the fundamental concerns of any sovereign state. Simultaneously, it encompasses multiple arenas within state-society relations, international relations, and civil-military relations, which are the primary loci for studying the autonomous power of the state. Lastly, as alluded to in the introduction and further elaborated upon in the subsequent section, the literature on space power exhibits a remarkable resemblance to the discourse on state power. In both realms, there is a predominant focus on capacity, highlighting the need for a thorough scrutiny of the concept of autonomy.

[22] Jonathan Hanson (2019) undertakes a deliberate endeavour to construct a "holistic framework" for measuring "state building and the growth of state capacity" by employing the State Capacity Dataset. While his theoretical motivations are persuasive and his methodological efforts commendable, we maintain that the overall undertaking is premature, particularly due to the omission of exploring the political dimension of autonomy and the presentation of the state solely as an amalgamation of "capacities."

2.3 Spacepower and Space Power[s]

The literature on spacepower/space power is limited if compared to the scholarly literature on *stateness*. It is also limited in comparison to other fields included in the fast-growing body of research on space affairs. The issue is that the concept is complex and sits across disciplinary boundaries. It does not fall perfectly under the umbrella of either space policy and politics, international (space) relations, or military theory, strategy and doctrine even though it relates to all three fields of research.

2.3.1 The Literature on Spacepower

The most striking observation about the existing literature on the topic is that it has been historically dominated by, and targeted at, the military sector. For most of its part, the literature has been indeed produced in a military-oriented milieu, with most notable scholars, analysts, and practitioners also being closely related to the US military complex in particular.[23] Over the last two decades, however, interest has been growing outside military circles, with the literature becoming increasingly multi-faceted and directed to wider scholarly and policy-making audiences. Chronologically, we segmented the existing literature in three distinct phases of development (see Fig. 2.5).

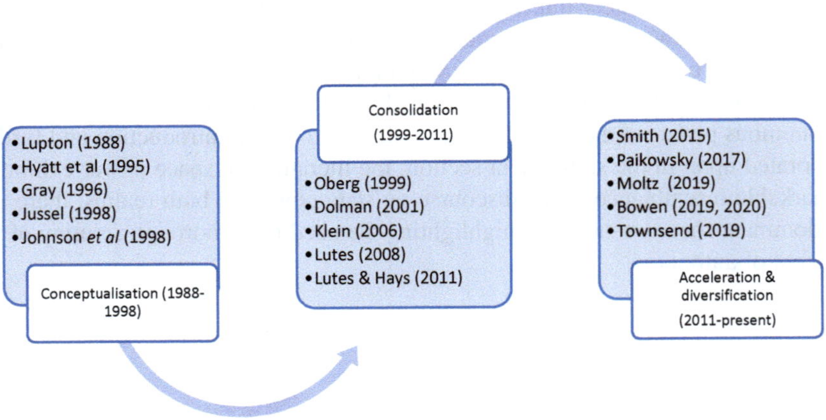

Fig. 2.5 Spacepower literature periodisation

---

[23] Prominent publishers on space power include Air University Press, Air and Space Power Journal, National Defence University Press, and Air Command and Staff College. Most research papers and theses accounting for much of the space power literature has been produced by or at one of these institutions.

The first, that we call the *conceptualisation* phase, opens with Lupton's (1988) early definition of spacepower as "the ability of a nation to exploit the space environment in pursuit of national goals and purposes [including] the entire astronautical capabilities of the nation" (Lupton, 1988: 4). Strongly informed by the 1991 First Gulf War, also known as the "first space war," these works focus on defining the concept in order to properly capture its complexity and operational utility. The conceptualisation effort continues at least until the aftermath of the Gulf War, when space capabilities clearly played a fundamental role in an armed conflict for the first time in history. Among the scholars having contributed to the early conceptualisation of spacepower figure (Billman 1997; Gray 1996; Hyatt et al., 1995; Johnson et al., 1998; Larned, 1994).

The second phase is clearly one of *consolidation*, with a number of seminal works including Oberg (1999), Gray and Sheldon (1999), Dolman (2001), Hays (2002) and Klein (2006). At this stage, the concept of space power/spacepower has become much more complex, with explicit reference to the combination between "technology, demographic, economic, industrial, military, national will, and other factors" (Oberg, 1999) and to the interdependence of space "with other media of warfare," including sea, land, or air (Klein, 2006).

In particular, James E. Oberg's *Space Power Theory* is widely regarded as the most comprehensive attempt at consolidating a theory of space power, as well as one of the milestones of the still-developing space power literature. Despite being published by a military institution, Oberg's *Space Power Theory* is not confined to the military domain. Indeed, Oberg has been arguably the first to provide a broad-ranging overview of space power features, by expanding the scope of the literature well beyond the defence domain. One of its major merits lies in the elaboration of what he calls the twelve "truths and beliefs" on space power,[24] which contributed to both theoretical and strategic aspects associated with shaping spacepower nationally and internationally.

Finally, the third and current phase of the debate sees the *acceleration* and *diversification* of these academic reflections, with more and more authors analysing space power/spacepower from different perspectives and with different approaches, including interpretative frameworks drawing from major IR schools of thought. This phase includes the works of Pfaltzgraff (2011), Kepron et al. (2011), M. Al-Rodhan (2012), Smith (2016), Paikowsky (2017), Czajkowski (2017), Bowen (2015, 2018,

[24] These include postulations such as: "Space exists as a distinct medium"; "Space power, alone, is insufficient to control the outcome of terrestrial conflict or ensure the attainment of terrestrial political objectives"; "Space power has developed, for the most part, without human presence in space, making it unique among other forms of national power"; "Technological competence is required to become a space power, and conversely, technological benefits are derived from being a space power"; "As with earthbound media, the weaponization of space is inevitable;" "At some time in the future, the physical presence of humans in space will be necessary to provide greater situational awareness;" "Situational awareness in space is a key to successful application of space power;" "Control of space is the linchpin upon which a nation's space power depends;" "Space operations have been and continue to be extremely capital intensive;" "Scientific research and exploration pays off;" and "There will be wild cards" (Oberg, 1999: 124-131).

2019, 2020), Moltz (2019) and Townsend (2019), among others (see Appendix A for an overview of the literature on spacepower between 1988 and 2020).

Notwithstanding this growing diversification, when looking at the definitions and theories proposed over time from an overall perspective, several recurring elements come to the fore.

A first striking observation about most definitions and theories circulating in the literature is that they are based on capacity-oriented conceptualisations of power (what Lupton referred to as *spacepower*). To date, the status of *space power* has received much less attention and has not been clearly defined, albeit a growing attention is paid to the division of space actors in tiers groups. Indeed, in a growing number of studies, space actors are classified into "tier groups." For instance, James G. Lee (1994), distinguishes *first-tier* space nations, possessing "dedicated military and civilian space capabilities on the cutting edge of technology" from *second-tier* nations which "develop and use dual-purpose space systems for both military and civilian purposes." *Third-tier* nations, in turn, are those that "lease or purchase space capabilities or products for military and civilian purposes from first and second tier nations" (Lee, 1994: 7). Other examples include Lambakis (2001), Newberry (2003), Harding (2012), Paikowsky (2017), Bowen (2018) and Klein (2019). Notably, this exercise has been mainly based on the measurement of their capacity, either military, civil or commercial.

A second common feature across most of the works is the focus on the military and defence agendas. Even if, to quote Lutes, "spacepower can be looked at through sociocultural, economic, and security lenses, each roughly equating to the civil-scientific, commercial, and military intelligence sector of space activity" (Lutes, 2008), theorists have so far looked at spacepower mostly through the military/security lenses. This is not surprising, given that the origin of the debates around spacepower was set in the context of the Cold War and the Gulf War, with the national security and military dimensions inevitably emerging as the predominant elements of spacepower definitions. In addition, until recently, spacepower theory has been almost exclusively a subject of discussion among military strategy scholars and personnel in military colleges and institutes. Furthermore, as contended by Bowen (2020: 42), spacepower theories must be necessarily oriented towards warfare—sometimes they might consider other aspects of space, but warfare is what strategic theories are for:

> [S]pacepower theory must itself not become involved in every aspect of space politics and development. Spacepower theory is about the instrumentalisation of violence with space technology; it is about war, not the entirety of relations between actors in space.

Some notable exceptions can be found. In addition, it is important to recognise that within the military realm of spacepower theories, different and competing "schools of doctrinal thoughts" stand out.[25]

[25] Lupton (1988) was the first author to provide an in-depth description of these schools, classified on the basis of their specific "belief structures" on four distinctive issues: a) the value of space forces, (b) the nature of space wars, (c) the employment of space forces, and (d) the latter overall organisational structure. The first school (Lupton, 1988: 22) was dubbed the "sanctuary doctrine," whose central belief was that space must be seen as a "war-free refuge." The second school was the "survivability

A third recurring, and rather popular element, in the military-oriented literature is the comparison of spacepower to land-, sea-, and airpower, with theorists trying to find common elements between space and other war-fighting domains. Here important lessons for the space medium are drown from classic works published in the XIX and XX century, including warfare theorist Carl von Clausewitz, seapower theorists Alfred Mahan and Julian Corbett, and airpower theorist Giulio Douhet. In other words, many theorists have referred to the "masters" and have drawn parallels with other prominent military models and domains (see Table 2.3).

This prevalent approach has been criticised, however, by a number of authors who have highlighted the possible flaws associated to recycling notions used for land-, sea-, and airpower, highlighting the uniqueness of the space medium instead. Oberg and to a lesser extent Dolman could be said to fall within this group. As pointed out by Hays (2002: 14), for instance, "[f]ew concepts from seapower theory translate directly into airpower theory, and it is not reasonable to expect either seapower or airpower theory to apply directly for the distinct medium of space." By trying to find useful analogies between land, sea, air and space, most theories tend to overlook the uniqueness of the space medium, which makes analogies technically hard or highly unlikely due to their implications. Oberg's *Space Power Theory* (1999: 43) makes this point with particular vigour when he notes:

> "Space power" is a phrase that evokes parallels with historical concepts of "sea power" and "air power." Useful parallels can be drawn. But without an appreciation for how different space is from air, sea, or land, false analogies and resulting erroneous decisions are possible, even likely.

A fourth recurring element is that most of the relevant literature bears an inherent prescriptive purpose and seeks to elaborate and recommend possible options for space policy decision-makers. This dimension is particularly evident in what Brian E. Fredriksson (2006: 45) calls the "propositional school" which includes theories developing around a finite number of propositions about space power features and implications.[26]

Within this widespread prescriptive approach, a specific leitmotiv relates to the value and role of space forces, their operative employment as well as their overall organisational structure, with a specific sub-theme being the urge to restructure such forces. Since the literature on spacepower is mostly US-based and imbued with references to the military sector, it is of no surprise that many spacepower experts have been advocating a reorganisation of the US space operation command, either within

doctrine," which also conceived space as a war-free domain but highlighted the intrinsic weakness of space assets (Lupton: 1988: 20). The third school was the "high ground doctrine," premised on "the old military axiom that domination of the high ground ensures domination of the lower lying areas" (Lupton: 1988: 21). The fourth school is the "control doctrine," which unlike high ground disciples, considered space forces as coequal with, and to some extent complementary to, air and sea ones.

[26] The other three schools identified by Fredriksson (2006) are: the "classical theorist school," where spacepower theories are somewhat extended from existing land-, sea-, and airpower theories; the "categorical school," which encompasses theories trying to divide the use of space in doctrinal categories.

Table 2.3 References to classic theorists in spacepower theories

Year	Author	Classic theorists referenced
1988	Lupton, David E.	Mahan, Mitchell, Clausewitz
1996	Gray, Colin S.	Clausewitz, Mahan
1997	Billman, Gregory	Douhet, Mitchell
1998	Jusell, Judson J.	Mahan, Douhet, Warden, Clausewitz
2000	France, Martin E. B.	Mahan
2001	Fox, John G.	Corbett
2001	Dolman, Everett C.	Mahan, Mackinder, Spykman
2003	Newberry, Robert D.	Mitchell, Douhet
2005	Wagner, John W.	Corbett, Mahan, Douhet, Mitchell, Clausewitz
2006	Fredriksson, Brian E.	Clausewitz, Mahan, Corbett, Douhet, Mitchell
2006	Klein, John J.	Corbett, Mahan, Douhet, Mitchell, Clausewitz
2006	Harter, Mark E.	Mitchell, Douhet
2008	Lutes, Charles D.	Mahan
2011	Sheldon, John B.; Gray, Colin S.	Clausewitz
2011	Smith, M. V.	Clausewitz
2011	Swilley, Scott F.	Mahan
2015	Ziarnick, Brent	Mahan, Clausewitz
2015	Bowen, Bleddyn E.	Mahan, Corbett, Clausewitz
2016	Smith, M. V.	Clausewitz
2019	Townsend, Brad	Clausewitz, Mahan, Corbett
2019	Moltz, James Clay	Mahan
2019	Bowen, Bleddyn E.	Mahan, Corbett
2019	Klein, John J.	Mahan, Corbett
2020	Bowen, Bleddyn E.	Mahan, Corbett, Clausewitz

the US Air Force or by the establishment of a separate military branch. The creation of an independent United States Space Force (USSF) is a case in point. The new structure was announced by President Donald Trump in June 2018 and was officially established the following year becoming the sixth armed service branch within the U.S. Armed Forces. Prior to its foundation, seminal works such as M. V. Smith's Ten propositions regarding Spacepower (2001) and M. E. Harter's Ten Propositions on Spacepower (2006) had urged for: (a) a separate theory, doctrine, and policy for space powers, framing space as a separate medium; (b) a class of professionals specialised in the space medium; and (c) a US Space Force which could embody these needs. The same had been stressed by Peter L. Hays et al. (2000), who co-edited an influential volume on *Spacepower for a New Millennium* and later contributed to the *Handbook of Space Security* with a chapter on Spacepower Theory (Hays, 2015). More recently, Hays (2020: 54) has updated his chapter to expand on the creation of the USSF and, in its *Spacepower Theory and Organizational Structures* he maintains that "organizational structures also play a critical role in shaping US spacepower."

> Focusing on the organizational structure of the Space Force and first-order priorities for space can help the United States ask the right questions and move toward doing the right things, at the right times, and for the right reasons.

Other recent works linking spacepower theory to the foundation of the USSF include Townsend (2019), Martindale and Deptula (2018) and Grosselin (2020). Overall, this widespread normative approach confirms that most military-oriented theories of spacepower are aimed not just at participating to the scholarly debate; they also (and perhaps primarily) aim at advising policymakers on how spacepower can—or should—be employed.

2.3.2 Recovering and Redefining Spacepower

Notwithstanding the rapid growth of the existing literature in the last 20 years and the emergence of many common elements, to date no holistic spacepower theory has yet emerged among scholars or space policy practitioners. We contend that a major explanation for this is that available theories provide no effective criteria to compare different countries. In particular, they lack a clear way to distinguish between space powers and lesser spacefaring actors. At the root of this difficulty lay two crucial weaknesses that should be analysed in more depth.

The first weakness is that the overwhelmingly capacity-oriented literature still tends to neglect the conceptual distinction between space power and spacepower—a distinction that was proposed three decades ago by David Lupton (1988: 4) and had promising implications both theoretically and methodologically. He argued convincingly that "spacepower is the ability of a nation to exploit the space environment in pursuit of national goals and that a nation with such capabilities can be termed a space power."

This crucial distinction has been lost in subsequent works, which generally use the two terms inconsistently, referring indistinctively both to more (or less) powerful actors and to the set of material (space) capacities that such actors possess—admittedly, in very different degrees. It is also important to highlight that the bulk of definitions and theories circulating in the literature has tended to focus on what, in Lupton's terminology, would be called *spacepower*, and to treat it as an array of material assets conferring a state the capacity to achieve first and foremost national security objectives. To illustrate, in 1994 Larned defined spacepower as "the ability to exploit the civil, commercial and national security space systems and associated infrastructure in support of national security strategy" (Larned, 1994: 4). One year later, spacepower was defined by Hyatt as "the ability of a state or non-state actors to achieve its goals and objectives in the presence of other actors on the world stage through control and exploitation of the space environment" (Hyatt et al., 1995: 5), while in 1996, Gray conceived spacepower as "the ability to use space for military, civil, or commercial purposes and to deny the ability of an enemy to do the same" (Gray, 1996: 299).

If in 1998 the Air Force Doctrine Document 2-2 defined spacepower as "the capability to exploit space forces to support national security strategy and achieve national security objectives" (USAF, 1998), the following year this definition became "the pursuit of national objectives through the medium of space and the use of space capabilities" (Johnson et al., 1998: 8). Quite relevantly, the latter authors also highlighted that "spacepower is connected to other forms of national power such as economic strength, scientific capabilities, and international leadership" (Johnson et al., 1998: 8). The latter point was more thoroughly addressed by James E. Oberg (1999), whose conceptualisation of spacepower arguably represents one the widest in scope and one of the clearest available, adding to the many reasons that make Oberg's work one of the most appreciated in this field. It defined space power as:

> the combination of technology, demographic, economic, industrial, military, national will, and other factors that contribute to the coercive and persuasive ability of a country to politically influence the actions of other states and other kinds of players, or to otherwise achieve national goals through space activity. (Oberg, 1999: 10)

Oberg's far-reaching definition features several elements which allow a nation to wield its spacepower. These so-called "elements of space power" are mostly rooted in the domestic arena or pertains to the structural conditions which the nations may face. They include: facilities, technology, industry, hardware and other products, economy, populace, education, tradition and intellectual climate, geography, and exclusivity of capabilities/knowledge. According to Oberg, "the more of these elements that are possessed by a user, the more flexible, reliable, and robust the applications will be" (Oberg, 1999: 44).

In a similar vein, more recent definitions connect spacepower to other forms of national power. For instance, Smith (2001: 6) underlined that "in the broadest sense, spacepower includes all activities performed by an actor—or exploited by an actor—in the space environment for civil, military, commercial, or other reasons." Lutes (2008: 67) argued that "spacepower both contributes to and is supported by other forms of power: diplomatic, informational, military, and economic, among others." More recently, Al-Rodhan (2012: 19) defined spacepower as the ability to use the space in function of what he calls *state capacities*, namely "social and health issues; domestic politics; economics; the environment; science and human potential; military and security; and international diplomacy," while Smith (2016: 160) once again highlights that "[s]pacepower is a vital element of a state's military instrument of power, but spacepower is also a vital element contributing to each instrument of a state's power: diplomacy, information, military, economic and culture."

Admittedly, the most recent literature on spacepower already conceives it in rather broad terms. In more recent years, in particular, there has been a visible expansion in the scope of the concept, with spacepower capabilities becoming intertwined with a nation's overall infrastructure, industrial, scientific, and institutional apparatuses, rather than being simply military- and national security-oriented. When "compared to earlier definitions, focusing almost exclusively on the military and national security realms, more recent definitions come with many different elements—economy, demography, industry, technology, etc.—and a wider spectrum of potential spacepower applications—pursuit of national goals, influence, persuasion, coercion—which finally hint at the all-inclusiveness of spacepower" (Aliberti et al., 2019: 8). Notably, this expansion is also reflected in the 2020 Doctrine for US Space Forces, which defines spacepower as "the totality of a nation's ability to exploit the space domain in pursuit of prosperity and security. National spacepower is comparatively assessed as the relative strength of a state's ability to leverage the space domain for *diplomatic, informational, military, and economic purposes*" (USSF, 2020: 13—emphasis added).

In all these definitions, spacepower overall emerges as an ability to use, exploit or control the space environment to pursue and achieve national goals, including first and foremost national security objectives. These conceptualisations juggle various recurring elements, alternating or combining them, but the primary emphasis revolves around the vital resources essential to assert spacepower. This focus on material strength is perhaps understandable, given the traditional prevalence of analysts of military, defence and security backgrounds. Hereby however, this strand of the literature ends up ignoring more complex, multidimensional conceptualisations of power. Significantly, they pass over the essential political component of sovereign decisional autonomy. Similarly, and consequentially, the concept of space power as a status—the quality of being actually powerful, not just strong—has received scant attention per se and has never been clearly defined.

Ultimately, a second major weakness of the existing literature on spacepower is strikingly similar to the one found in studies on state power: it mainly focuses on capacity and largely ignores autonomy. The following two sections tackle this issue in detail by reframing these two dimensions in relation to spacepower.

2.3.2.1 Reframing Capacity

As noted above, most definitions and theories circulating in the literature are based almost exclusively on capacity-oriented conceptualisations of power. Their overall focus is on the set of material capacities required to acquire and exercise spacepower and thus (rather confusedly) to "be" a space power. In this sense, all states capable to act in space are by definition "space powers" independently of how much *spacepower* they actually muster—it is just a matter of degree.

Capacity, in sum, is turned into an overstretched umbrella notion, a "tyrannical" concept that covers everything that is needed to set and achieve national goals, including first and foremost national security objectives.

Hypertrophic as it is, however, the concept of capacity in space literature is also incomplete. Capacity-driven studies in fact have been focusing almost exclusively on what we would term "hard" capacity, i.e., resources, expertise and technical skills required to build, own, and operate space assets. One reason this is the case is that hard capacity is relatively easier to measure as it deals first and foremost with objective indicators.

The concept of capacity thus is in need of a comprehensive reframing. This can be done in two ways.

First, it should be reframed by further enlarging it to include a dimension we call "soft" capacity. This focuses on social skills and governance activities and aims to assess to what extent space is effectively embedded in the national infrastructure and leveraged to implement national policies in various fields, including—beyond security, defence and diplomacy matters—social welfare, the environment, sustainable development, etc.

Second, we contend that to better assess the capacity of a state to "use space," these two subdimensions (hard capacity and soft capacity) should be first separately assessed and then integrated into a single measure.

In our conceptualisation, *hard capacity* relates to the material assets and abilities that enable an actor to operate in, through and from space in the full spectrum of space activities. *Soft capacity* denotes instead the ability to effectively utilise and integrate assets and expertise in national policies, infrastructure, and activities.

As shown above, capacity-driven studies of space affairs have been focusing almost exclusively on the first, "hard" sub-dimension, as it encompasses all the capabilities required to build, own, and operate space assets. Also, hard capacity deals first and foremost with objective indicators (e.g., the number of launches operated by a certain country, or the number of satellites owned). Conversely, "soft capacity" measures to what extent space is effectively embedded in the national infrastructure and leveraged to implement national policies in various fields, including among others—and besides the obvious cases of the defence and security sectors—the economy, the environment, welfare, education, entertainment, science, etc.

In addition to reconceptualizing capacity in the form of hard capacity and soft capacity, it is also fundamental to ultimately scale it back to make room for its neglected twin concept: autonomy. It should be recognised that, in order to grasp the whole content and meaning of "space power," capacity and autonomy should be used, measured, and evaluated in conjunction.

2.3.2.2 Incorporating Autonomy

As said above, the mainstream space literature fails to pay specific attention to the concept of autonomy. It does incorporate, however, a "hard" or "technical" version of the concept, which refers to the ability of a state to ensure mission independence in case of necessity vis-s-vis foreign actors.

To illustrate, in a volume edited by the European Space Policy Institute (Al-Ekabi, 2015a) specifically dedicated to addressing the complex issue of autonomy in space, Densing and Reinke (2015) discusses the need for ensuring European independence in space applications, particularly in satellite navigation, Earth observation, satellite communications and space situational awareness. De Winne (2015) presents the challenges associated to the pursuit of autonomy in human spaceflight, specifically with regard to the development of indigenous human-rated launchers and capsules. Al-Ekabi (2015b) identifies the factors that favour or prevent autonomy in space transportation and discusses the potential costs of a non-autonomous access to space for Europe, while Tortora (2015), discusses the longstanding issue and negative implications of European technological dependence. He focuses in particular On Europe's need to outsource certain components, raw and advanced materials as well as some basic technologies and building blocks that are not available within European boundaries. In a more recent study, Fiott (2020: 18) defines a strategically autonomous actor as "one that can design, develop, launch and operate space systems without hindrance," highlighting that today the EU space programmes, "Galileo, EGNOS and Copernicus offer the EU a high level of strategic autonomy [that] allows it to observe earth, protect transport networks, sustain digital networks and the security of trade routes and much more."

In all these studies, the concept of autonomy is roughly equivalent to that of "self-sufficiency," "technological independence" or "non-dependence,"[27] as it entails a preference for domestic solutions and the achievement of strategic independence (above all energetic and technological) in the international arena.[28]Empirically, self-sufficiency can be measured by hard data. For instance, ASD-Eurospace (2020) has defined and reviewed a list of critical technologies to be developed by the European space sector to foster non-dependence in space, estimating the economic costs associated with the effort. This has brought some to conflate self-sufficiency with hard capacity, irrespective of the conceptual distinction between the two. For this reason we prefer to isolate these aspects and label them "hard autonomy." Given the absence of a comprehensive comparative measurement of this particular aspect of state autonomy in space, we present here the first attempt in this direction.

As for "soft" or "political" autonomy proper, the concept is to be introduced from scratch in space studies. In the space context, political autonomy means that political decision makers at the helm of the state (be them elected elites or autocratic rulers) are able to decide on space matters in the "national interest" however they may define it. This entails being independent from "coercions or impositions" coming from competing interests or alternative centres of power, whether located abroad or at home.

Foreign agents of influence include first and foremost other states but may also extend to international institutions and multinational corporations. Domestic agents

---

[27] As explained by Caito (2015), "technological independence" implies that all needed space technologies are developed inside a country, while "technological non-dependence" refers to the possibility for an actor to have free, unrestricted access to any required space technology.

[28] The conceptual difference between "autonomy" and "strategic independence" was interestingly explained by Jan Wouters and Rik Hansen (2015) in the book European Autonomy in Space: "autonomy is a political concept by nature and can be defined as'possessing the power to determine one's own laws'. Strategic independence is more operational and refers to'the capacity to take the required decisions and to execute them so as to safeguard a number of vital interests'. Autonomy could thus be understood as a formal criterion (the goal to achieve), while strategic independence is the necessary condition for effective autonomy (the means to reach the goal)" (Al-Ekabi, 2015b).

of influence, in turn, would include actors that either lay outside the public sphere (as with large private corporations trying to influence or even "capture" the policy process), or are located within the state apparatus, such as the military, technocrats and managerial elites of state-owned corporations.

The space sector offers numerous examples highlighting the challenges these actors can pose to the political autonomy of a state. A paradigmatic case of foreign influence in the space sector is represented by the 1990 US-Japan Satellite Procurement Agreement, which ended Japanese protection of its still developing satellite market from international tendering. The agreement finds its root in the mounting trade frictions between Japan and the United States during the 1980s. Because the US saw a possible threat coming from the nascent Japanese satellite market, it activated the so-called Super 301 as part of their trade talks over semiconductors in the mid-1980s (Aoki, 2009; Dunphy, 2016). As explained by Office of the US Trade Representative (2000), Section 301 of the Trade Act of 1974 is "the principal U.S. statute for addressing foreign unfair practices affecting U.S. exports of goods or services. Section 301 may be used to enforce U.S. rights under international trade agreements and may also be used to respond to unreasonable, unjustifiable, or discriminatory foreign government practices that burden or restrict U.S. commerce." "Super 301," in turns refers to an annual process by which the US Trade Representative identifies those unfair traders, engage in trade negotiations and, in case unfair trade practices are not eliminated, impose sanctions. As reported by Dunphy (2016: 30), "Japan was named an unfair trading nation in 1989, and negotiations began on forest products, supercomputers, and telecommunications satellites." Unlike other economic sectors, at the time the Japanese government did not deem space as a strategically vital sector, and eventually surrendered to the American claims, agreeing on the signature of the Japan Satellite Procurement Agreement in 1990 (Zeng, 2004). The Procurement Agreement required Japan to open the procurement of its commercial satellites, particularly telecommunications satellites, to foreign satellite manufacturers through tendering. The only exclusion to this provision was "R&D satellites, defined as satellites designed and used entirely, or almost entirely, for the purpose of in-space development and/or validation of technologies new to either country, and/or non-commercial scientific research" (Dunphy, 2016). According to Setsuko Aoki (2009), "that provision was tantamount to a death sentence to the embryonic Japanese satellite industry," as it ultimately rendered Japanese technology—which at that time was characterised by high cost due to small government orders—non-competitive vis-à-vis US-manufactured satellites. US satellite manufacturers, which already had economies of scale to offer lower cost satellites, entered into the Japanese market. Still today, the vast majority of communication satellites owned by Japanese telecommunication companies are US-made (Aliberti & Hadley, 2020). This episode, among other things, provides an example of the connection between soft and hard autonomy. Political pressures from the USA forced Japan to reduce its self-sufficiency in relation to the American state and industry. The target was not Japan's space capacity, but rather its technical and, ultimately, political autonomy from the US.

Influence on policy making is not necessarily exercised by foreign states only. As mentioned above and discussed in detail in Sect. 2.2.2, it can also be exercised by domestic and multinational corporations, which can affect internal and external decision-making processes alike. Indeed, if autonomy is best seen as synonymous with sovereign decisions, then strong private interests may under some circumstances limit the ability of states to independently pursue their strategic goals by influencing or even "capturing" the policy process. A well documented example in this regard is offered by the aggressive lobbying implemented by the US-based multinational Motorola Inc. in the early 1990s to secure favourable decisions in the allocation of radio frequencies. Specifically, the company, which in the early 1990s had financed the development of the first constellation of Iridium satellites, lobbied the national delegations of the World Radiocommunication Conference 1992 to have the conference decide for frequencies allocations favourable to the deployment of the Iridium constellation. As documented by John Bloom (2016), this was a huge and well-prepared capture effort, with Motorola successfully obtaining the craved frequency allotment. The move however also caused the wrath of some European nations and telecommunication companies. This is not to mean that a vibrant private sector pursuing its legitimate business interests is per se an indicator of state weakness. The issue is not even whether private companies wield political influence, but to what extent they exercise it to favour private interests at the expense of the public good. And in the case of a multinational corporation, the notion of public good may well extend beyond the boundaries of the country where it originated.

Besides private companies, state-owned enterprises (SOEs) can influence policy-making as well. Although they are an essential component of the state and their leaders are part of a bureaucratic-technocratic establishment, they may develop interests and priorities of their own acting as agents of influence from within the state. State-owned enterprises (SOEs) can influence government policy decisions in the space realm in a number of ways. Two examples of this are Roscosmos and China Aerospace Science and Technology Corporation (CASC). These SOEs both have a monopoly on the launch of satellites in Russia and China, respectively, which gives them a great deal of influence over their governments' space policy. For example, Roscosmos has been able to block the launch of foreign satellites that it deems to be a threat to Russian national security. CASC, on the other hand, has been able to block the development of private space companies in China for a long time. In both cases, there were strategic and policy disagreements coming both from within the state and from the private sector. These instances raise still unanswered questions about the actual making of space policy in these countries, the relationship among different sectors of the political elites, the managerial-technocratic establishments, and private business leaders.

As for the influence exercised by the military, an interesting case is offered by the change in the policy posture of the Obama administration on space security issues, which was stirred by the US defence community. During its first term (2008–2012), the Obama administration enacted a National Space Policy (NSP) in 2010 and a National Security Space Strategy (NSSS) in 2011. Through these documents, the administration changed the posture held by the Bush administration and redirected the United States towards a cooperative, civilian and commercial-oriented approach to space activities. In the area of space security, the "strategic restraint" of the Clinton era was favoured, whereby the US refrained from deploying offensive counterspace capabilities with a view to moderate the behaviour of both friends and potential adversaries (Hitchens & Johnson-Freese, 2016). All this suddenly changed after May 2013, when China launched what the US military deemed an anti-satellite weapons test that nearly reached geostationary orbit. As reported by Teresa Hitches and Joan-Johnson Freese (2016: iii):

> The Chinese test, coming on the heels of both Russia and China testing manoeuvrable satellites in low Earth orbit—a capability that, until recently, had been demonstrated only by the United States—led to something of a "quiet panic" within the US national space security community. This renewed threat perception, and the renewed fear about the "inevitability" of space war, was elevated all the way to President Obama (which is somewhat rare in the space strategy world), triggering a summer 2014 National Security Council-led Strategic Portfolio Review. Consequent to that exercise, changes in force posture, development programs, and budgets have been initiated to forge a more muscular national security space strategy [...] Defense against counterspace capabilities has taken on a top priority, followed by in order, a diminished view of space diplomacy, and an increased interest in offensive capabilities. In particular, the increased threat perception was accompanied by more aggressive public diplomacy by the Pentagon and US Air Force, aimed at making it very clear that the United States would respond to threats in space with the use of force—with rhetoric slipping back toward the "dominance and control" motif of the Bush administration's space policy.

All the examples mentioned above serve as notable occurrences showing how political autonomy in space can be challenged. The issue now is to go beyond anecdotic narratives and find a way to empirically assess the level of political autonomy of different space actors and integrate the data in a comprehensive comparative measure of "state autonomy in space."

Ultimately, we posit that *technical autonomy* (or *hard autonomy*) relates to the state ability to access and operate in space without relying on external sources of supply. *Political autonomy* (or *soft autonomy*) is defined instead as sovereignty over space matters, and concerns the state's ability to independently formulate space-related policies regardless of conflicting *particular* interests that may arise internally or externally. It goes without saying that technical autonomy does not imply full-fledged autarchy, but rather the possibility to ensure mission independence in case of necessity (e.g., in case of an impending natural disaster, a trade dispute, or a war). The notion of technical autonomy entails a preference for domestic solutions over foreign ones and the achievement of strategic (above all energetic and technological) independence from external actors: a fully autonomous power is not necessarily an "isolationist" one—in fact, full-blown autarchy may even prove detrimental –, but rather a nation which can rely on diverse sources of supplies.

As for "political autonomy," the concept can be further broken down into two components. The first is *external political autonomy*, which comprises the notions of international legal, Westphalian and interdependence sovereignty popularised by Krasner (1999: 3–42). In the space context, external political autonomy can be understood to refer to the state's ability to determine national goals without being influenced by invitations and interventions (i.e., coercions or impositions) by foreign actors. The second subdimension is *domestic political autonomy*, and relates to a state's decision-making independence. It requires that political decision-makers at the helm of the state (be them elected elites or autocratic rulers) be able to decide on space matters in the 'national interest'—however they may define it—against the opposition of competing or alternative centres of power. The latter would include actors whose interests either (a) lay squarely outside the public sphere, such as private companies); or (b) are rooted inside the state apparatus, such as the military, bureaucratic and technocratic elites including top-level managers of state-owned enterprises.

Once the two fundamental dimensions of capacity and autonomy are clearly defined, they can be used in conjunction to determine the power status not only of space powers, but also of different space actors.

2.3.3 *Spacepower as a Matrix:* Space Powers *and Other Actors*

Mainstream literature does recognise the need if not to measure spacepower, at least to discriminate among states that possess different amounts of space capabilities. Many studies indeed have tried to frame space actors into tier groups (e.g., Bowen, 2018; Harding, 2012; Lambakis, 2001; Newberry, 2003). Not surprisingly, however, by conflating all these "capacities" into a single unidimensional measure, the graphical representation of space power tiers can only be a column or a pyramid. And, due to the simplistic and unidimensional nature of the distinction between tiers, the categorisation of "space powers" can only be delineated by using adjectives such as higher/lower, or small/medium/great, or primary/secondary/tertiary, and the like. A poignant illustration of this is the pyramidal structure of the "space club" elaborated by Paikowsky (2017). According to this author, the difference between space powers—i.e., what makes countries climb or fall between tiers—lies simply on the accumulation of material assets and other resources that enable them to achieve specific goals and implement specific policies. In other words, what is being measured here is a country's *capacity* to reach and act in outerspace, which some states possess in larger measure than others.

Given our understanding of power as a status that encompasses not only the material capacity to act but also, and primarily, the ability to autonomously determine one's objectives and lines of action, traditional representations such as pyramids and columns prove inadequate in capturing the complexity of reality. These linear structures fail to reflect the multidimensional nature of power and the interconnectedness of its components. As illustrated above in general terms (see Figure 2.4), a matrix provides instead the most suitable graphic representation for our framework. It visually illustrates that "space power" occupies a specific area in the Cartesian plane. It also highlights that in order to be recognised as a space power, an actor must exceed multiple thresholds along the axes of capacity and autonomy, rather than just one. The matrix effectively conveys the multidimensional nature of power in the space domain and reveals that only a space actor with both high capacity and high autonomy can be classified as a space power. It also demonstrates that various

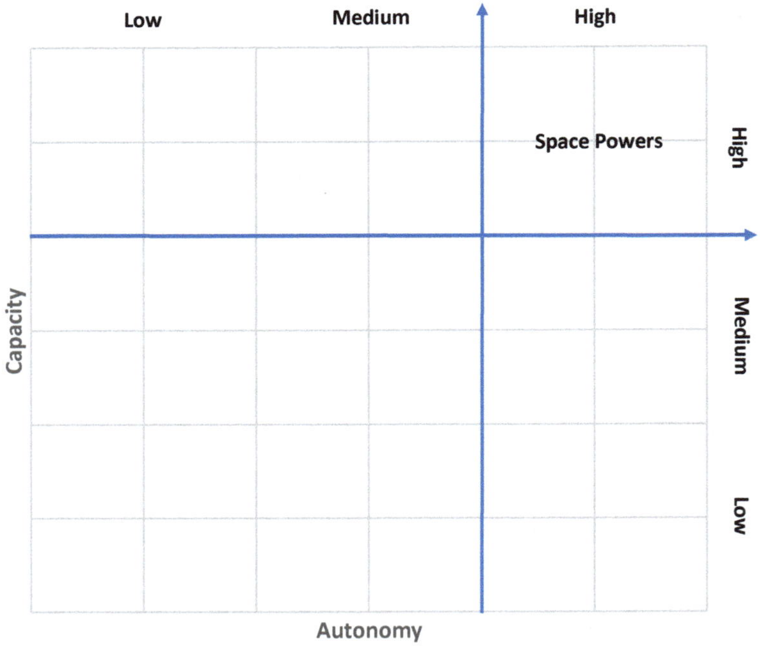

Fig. 2.6 A conceptual spacepower matrix

trajectories exist for entering or exiting the space power area, depending on an actor's position at a given point in time and the specific components of power (capacity or autonomy) that require enhancement or risk reduction (Fig. 2.6).

Figure 2.2 shows a graphic representation of our spacepower matrix. If only space actors that present a combination of high levels of capacity and high levels of autonomy can be considered space powers, then what are the space actors that fall short of this definition in one or multiple counts? In order to answer this question, we turn our attention now to providing a new definition of the core concept of spacepower and, subsequently, devising a method to identify and measure the various dimensions in which it can be articulated.

Shifting the focus from a purely unidimensional conceptualisation to one entailing different dimensions first raises a major definitional task. Our definition of space power is provided in Box 2.1.

Box 2.1 Definition of Space Power

An entity endowed with a high degree of *spacepower*, i.e., possessing the means to deploy, operate, and benefit from space-related capabilities as well as the ability to pursue national objectives that it has autonomously determined.

This definition presents five substantial advantages over other conceptualisation *in vogue* in the scientific literature. Firstly, it allows for conceiving *space power* not as an attribute inherent to all states, albeit to varying degrees, but rather as a status which can only be acquired and maintained by meeting specific requirements. Space power status is roughly equivalent to great power status, though albeit confined to the space domain: it is a "positional" concept and implies a division of actors into typological groups, rather than hierarchical tiers.

Secondly, it is consistent with the distinction between space power and *spacepower* as first outlined by Lupton (1988) but disregarded by subsequent authors: *spacepower* consists of the set of attributes which any actor must possess in a sufficient degree in order to be considered a *space power*. This distinction has a major impact on the measurement model: indeed, while spacepower is best seen as a "continuous" variable, space power is only a "type" of a broader taxonomy represented as a matrix.

Thirdly, it enables us to move from a unidimensional to a multidimensional model. To empirically measure spacepower and identify space powers, we need to consider the power status of an actor as the intersection of both its capacity (the ability to do things: its *skills*) and its autonomy (the ability to independently determine what things are to be done: its *self-reliance*).

Fourthly, although it does not explicitly rule out the possibility that non-state, private actors may be endowed with some attributes of *spacepower*, our definition recognises that, for the time being, only nation states have the means and the ability and the legitimacy to master the entire spectrum of spacepower components and, thus, claim the status of space power. In particular, private actors cannot achieve full autonomy, as they lack the distinctive attributes defining modern sovereignty and must abide by state laws and regulations. Space power, in other words, is a form of state power.

Lastly, our definition leaves the door open to the complexity of the matter at hand. In fact, as we have seen, capacity and autonomy can be further broken down into four constituent elements or "sub-dimensions": *hard* and *soft capacity*, and *technical* and *political autonomy,* respectively. This approach allows for the identification of space powers across different time periods and technological contexts, as the multidimensional concept of *spacepower* and its constituent elements adapt to reflect key developments in available technologies, the ever-expanding scope of space activities, and the changing distribution of technical and political abilities in both the domestic and the international arena.

As shown in Fig. 2.7, only a country with a combination of high capacity and high autonomy is a space power; countries sporting combinations of low capacity and low autonomy are not going to be prominent players in the space arena. However, while in the 1960s and 1970s only a handful of countries were actively involved in space activities, nowadays virtually all countries engage in a variety of ways with space and space technology. As matter of fact, at the time of this writing, 46 national and regional space agencies are listed in the website of the United Nations Office for Outer Space Affairs. This number climbs to nearly 80 if we add space agencies that have been announced or have only been formally established but are still in the process of being properly set up. Since engagement with outer space seems to be something that most countries nowadays aspire to achieve, we define countries that currently present a combination of low levels of capacity and low levels of autonomy as "emerging space nations." The group includes countries that are not active in space, as well as countries that have only marginal engagement with the space domain.

Defining the two extremes of our matrix leaves a large area in the middle that is currently undefined. This area includes all countries that the existing literature defines as "spacefaring nations." In our framework, a spacefaring nation has achieved a certain level of capacity and/or autonomy that allows it to engage with outer space, albeit not at the level of a space power. Figure 2.7 graphically represents the three typologies discussed.

In many ways, the spacefaring nations group is the most interesting of the three. We fully expect that countries with enormous space programmes such as the United States, China, and Russia will emerge from our measurement as space powers, and it seems unlikely that they will lose this status any time soon. We also would expect a country such as Rwanda—which is not part of this study—to fall squarely in the "emerging space nations" group, with its very ambitious space programme recently announced, and we don't expect it to become a space power any time soon. On the contrary, spacefaring nations may have enough spacepower, in terms of both capacity and autonomy, to aspire to space power status. So far, however, in the

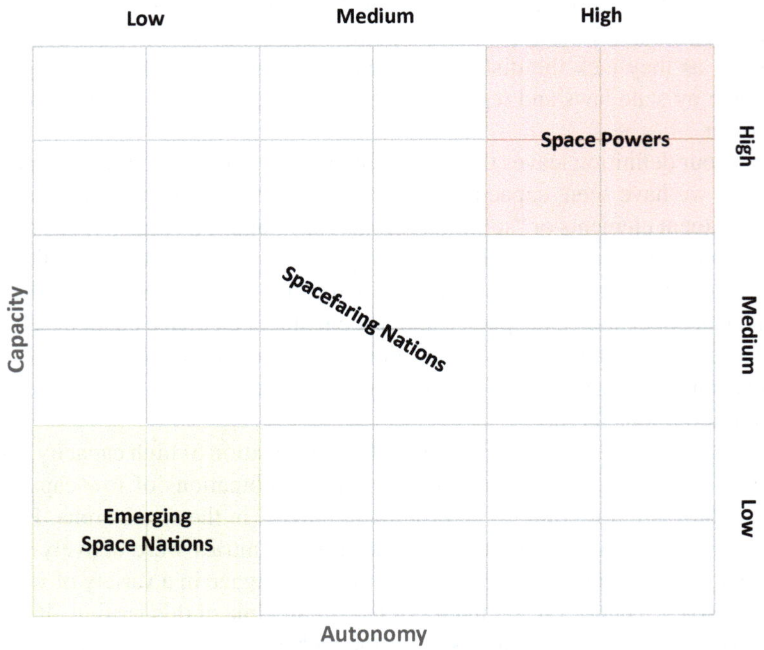

Fig. 2.7 Space powers, emerging space nations, and spacefaring nations

existing literature (e.g., Bowen, 2018; Harding, 2012; Lambakis, 2001; Newberry, 2003) scholars have usually divided space actors into three tiers and ultimately this middle tier of "spacefaring nations" serves as a catch-all category for countries that do not qualify as space powers but still have some engagement with outer space beyond minimal levels.

Reality, however, is more complex than that. In fact, while increasing autonomy and capacity, it is very likely that different states pursue different strategies. Even more interestingly, there may be some level of trade-off between capacity and autonomy, in the sense that it may be possible for a state, for instance, to increase its capacity while limiting its autonomy. As a consequence, the "spacefaring nations" category needs a better conceptualisation. We propose to divide this "middle tier" in three distinct groups. The first, which we define "skilled spacefaring nations," include countries that have reached a certain degree of capacity but do not perform well in the autonomy dimension. Similarly, the second group includes states where substantial levels of autonomy are associated with low levels of capacity and is defined "self-reliant spacefaring nations." Finally, the third group includes states that present a certain *quantum* of both capacity and autonomy and is defined "primed spacefaring nations": they are on their way to achieve the coveted status of space powers and possess the raw material needed to do so in terms of a successful combination of both autonomy and capacity. Figure 2.8 graphically represents all 5 proposed categories within the spacepower matrix.

Conceptualizing the boundaries shown in Fig. 2.4 is not an easy task. We choose to add some nuance to the categories by setting thresholds that take into account both the level of each dimension and the balance between the levels of both dimensions. In that sense, we allow room for the possibility that a country sporting very high levels of capacity be still considered a space power even if it shows relatively lower levels of autonomy, or vice-versa, provided that the level reached in the lower dimension is still considerable (i.e., medium–high). We do acknowledge, however, that these boundaries can be somewhat arbitrary, and that this framework should be interpreted more with a focus on movement towards the 5 ideal-types located in the corners and centre of the matrix, rather than aiming to the transition from one arbitrary boundary to another.

Based on the conceptual framework described thus far, in the following chapters we will introduce our measurement model and present the empirical analysis of the data collected for the 11 countries under examination.

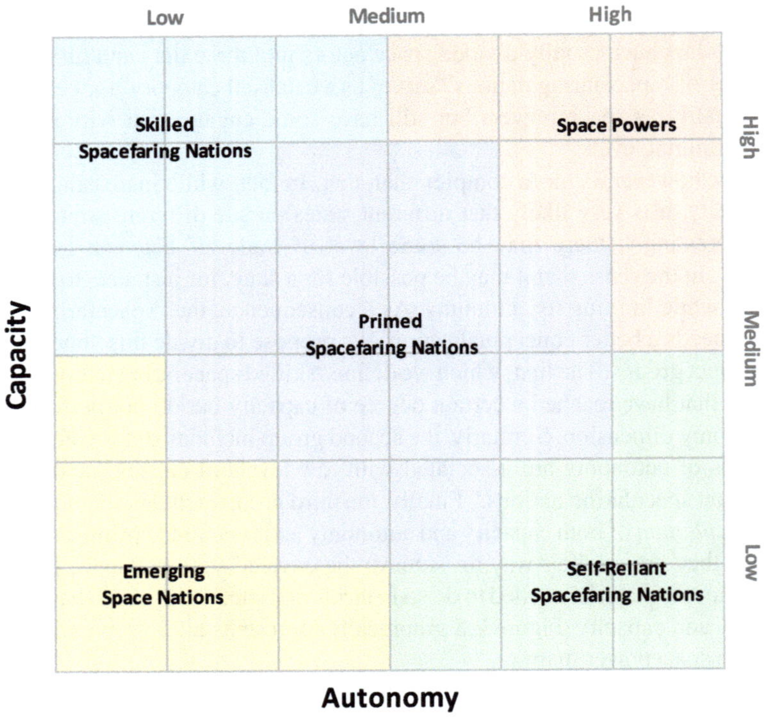

Fig. 2.8 The conceptual spacepower matrix

References

Acemoglu, D., & Robinson, J. (2020). *The Narrow Corridor: States, Societies, and the Fate of Liberty*. Penguin Press.

Afanas'ev, M. (2004). Nevynosimaya slabost' gosudarstva. *Otechestvennye zapiski* (vol. 2).

Afanas'ev, M. (2009). The quality of the state—Russia's chief problem. *Russian Politics & Law, 47*(4).

Al-Ekabi, C. (ed.). (2015a). *European Autonomy in Space*. Springer.

Al-Ekabi, C. (2015b). European Access to Space: Factors of Autonomy. In C. Al-Ekabi (Ed.), *European Autonomy in Space*. Springer.

Al-Rodhan, N. R. (2012). *Meta-Geopolitics of Outer Space*. Palgrave Macmillan.

Alesina, A., & Dollar, D. (2000). Who Gives Foreign Aid to Whom and Why? *Journal of Economic Growth, 5*(1), 33–63.

Alesina, A., & Weder, B. (2002). Do Corrupt Governments Receive Less Foreign Aid? *American Economic Review, 92*(4), 1126–1137.

Aliberti, M., & Hadley, S. (2020). *Securing Japan. An assessment of Japan's Strategy for Space* (ESPI Report 74). European Space Policy Institute.

Aliberti, M., Cappella, M., & Hrozensky, T. (2019). *Measuring Space Power: A Theoretical and Empirical Investigation on Europe*. Springer.

Aoki, S. (2009). Current Status and Recent Developments in Japan's National Space Law and its Relevance to Pacific Rim Space Law and Activities. *Journal of Space Law, 35*, 363–438.

APSA. (2004). *American Democracy in an Age of Rising Inequality*. American Political Science Association, Task Force on Inequality and American Democracy.
ASD-Eurospace. (2020). *Space RDT Priorities. The Incremental Roadmap of Technology Research and Development Activities for Space*. ASD-Eurospace. https://eurospace.org/wp-content/uploads/2018/05/eurospace-rdt_2020_web.pdf (Accessed on 20 November 2022).
Badie, B., & Birnbaum, P. (1983). *The Sociology of the State*. The University of Chicago Press.
Barkey, K., & Parikh, S. (1991). Comparative Perspectives on the State. *International Review of Sociology, 17*(2), 523–549.
Bartel, A. P., & Thomas, L. G. (1987). Predation Through Regulation. *Journal of Law & Economics, XXX*.
Baru, S. (2009). The Influence of Business and Media on Indian Foreign Policy. *India Review, 8*(3), 1–23.
Bayne, N., & Woolcock, S. (Eds). (2017). *The New Economic Diplomacy: Decision-Making and Negotiation In International Economic Relations*. Routledge.
Billing, T. K., & Farro, A. (2016). Cultural Correlates of Crony Capitalism in India. In N. Khatri, & A. K. Ojha (Eds). *Crony Capitalism in India. Establishing Robust Counteractive Institutional Frameworks*. Palgrave.
Billman, G. (1997). *The Inherent Limitations of Spacepower: Fact or Fiction?* Air University Press.
Bloom, J. (2016). *Eccentric Orbits: The Iridium Story*. Atlantic Books.
Bolton, J. R. (2001). The Risks and Weaknesses of the International Criminal Court from America's Perspective. *Law and Contemporary Problems, 64*(1).
Bosco, D. (2014). *Rough Justice. The International Criminal Court's Battle to Fix the World, One Prosecution at a Time*. Oxford University Press.
Bowen, B. E. (2015). *Spacepower and Space Warfare: The Continuation of Terran Politics by Other Means*. Aberystwyth University.
Bowen, B. E. (2018). British Strategy and Outer Space: A Missing Link? *The British Journal of Politics and International Relations, 20*(2), 323–340.
Bowen, B. E. (2019). From the Sea to Outer Space: The Command of Space as the Foundation of Spacepower Theory. *Journal of Strategic Studies, 42*(3–4), 532–556.
Bowen, B. E. (2020). *War in Space: Strategy, Spacepower*. Edinburgh University Press.
Bueno de Mesquita, B., & Smith, A. (2007). Foreign Aid and Policy Concessions. *Journal of Conflict Resolution, 51*(2), 251–284.
Caito, L. (2015). *European Technological Non-dependence in Space* (ESPI Report 51). European Space Policy Institute.
Carpenter, D., & Moss, D. (Eds). (2014). *Preventing Regulatory Capture*. Cambridge University Press.
Carter, D. B., & Stone, R. W. (2015). Multilateralism and Democracy: The Case of Vote Buying in the United Nations General Assembly. *International Organization, 68*(1), 1–33.
Che, Y. (1995). Revolving Doors and the Optimal Tolerance for Agency Collusion. *RAND Journal of Economics, 26*(3) 378–397.
Cingolani L. (2013). *The State of State Capacity: A Review of Concepts, Evidence and Measures* (MERIT Working Papers 2013–053). United Nations University—Maastricht Economic and Social Research Institute on Innovation and Technology (MERIT).
Cingolani, L. (2018). The Role of State Capacity in Development Studies. *Journal of Development Perspectives, 2*(1–2), 88–114.
Coppedge, M., et al. (2020). *V-Dem Codebook v10. Varieties of Democracy (V-Dem) Project*. University of Gothenburg.
Czajkowski, M. (2017). Theory of Spacepower—A Brief Introduction. *Jagielloński Przegląd Bezpieczeństwa, 2*(3), 25–51.
Dahl, R. A. (1961). *Who Governs?* Democracy and Power in an American City. Yale University Press.
Dahl, R. A. (1971). *Polyarchy: Participation and Opposition*. Yale University Press.
Dahl, R. A. (1989). *Democracy and Its Critics*. Yale University Press.

dal Bó, E. (2006). Regulatory Capture: A Review. *Oxford Review of Economic Policy, 22*, 203–225.
Darden, K. (2008). The Integrity of Corrupt States: Graft as an Informal State Institution. *Politics and Society, 36*(1), 35–60.
De Winne, F. (2015). European Autonomy in Space: Human Space Flight. In C. Al-Ekabi (Ed.). *European Autonomy in Space*. Springer.
Densing, R., & Reinke, N. (2015). The Need for European Independence in Space Applications. In C. Al-Ekabi (Ed.). *European Autonomy in Space*. Springer.
Dolman, E. C. (2001). *Astropolitik: Classical Geopolitics in the Space Age*. Routledge.
Dowding, K. M. (Ed.). (2011). *Encyclopaedia of Power*. SAGE.
Downs, A. (1957). *An Economic Theory of Democracy*. Harper.
Dreher, A., Sturm, J. E., & Vreeland, J. R. (2009a). Development Aid and International Politics: Does Membership on the UN Security Council Influence World Bank Decisions? *Journal of Development Economics., 88*(1), 1–18.
Dreher, A., Sturm, J.-E., & Vreeland, J. R. (2009b). Global Horse Trading: IMF Loans for Votes in the United Nations Security Council. *European Economic Review, 53*(7), 742–757.
Duncan, P. (2007). *'Oligarchs', Business and Russian Foreign Policy: From El'tsin to Putin* (Economics Working Papers 83). Centre for the Study of Economic and Social Change in Europe, SSEES. University College London.
Dunphy, R. (2016). *Space Industry Business Opportunities in Japan: Analysis on the Market Potential for EU SMEs Involved in the Earth-Observation Products & Services*. EU-Japan Centre for Industrial Cooperation.
Easter, G. M. (2008). The Russian State in the Time of Putin. *Post-Soviet Affairs., 24*(3), 199–230.
Eldar, O. (2008). Vote-trading in International Institutions. *The European Journal of International Law, 19*(1).
Enriquez, E., & Centeno, M. A. (2012). State Capacity: Utilization, Durability, and the Role of Wealth vs. History. *International and Multidisciplinary Journal of Social Sciences, 1*(2), 130–162.
Evans P. B., Rueschemeyer, D., & Skocpol, T. (Eds). (1985). *Bringing the State Back In*. Cambridge University Press.
Evans, P. (1995). *Embedded Autonomy: States and Industrial Transformation*. Princeton University Press.
Feaver, P. D. (2003). *Armed Servants: Agency, Oversight, and Civil-Military Relations*. Harvard University Press.
Finer, S. E. (1997). *The History of Government. Vol III. Empires, Monarchies and the Modern State*. Oxford University Press.
Fiott, D. (2020). *The European Space Sector as an enabler of EU strategic autonomy*. European Parliament. EP/EXPO/A/SEDE/FWC/2017-01/05.
Fortescue, S. (2006). *Russia's Oil Barons and Metal Magnates: Oligarchs and State in Transition*. Macmillan.
Fredriksson, B. E. (2006). *Globalness: Toward a Space Power Theory*. Air University Press.
Fukuyama, F. (1992). *The End of History and the Last Man*. Free Press.
Fukuyama, F. (2004). *State-Building: Governance and World Order in the 21st Century*. Cornell University Press.
Fukuyama, F. (2014). *Political Order and Political Decay: From the Industrial Revolution to the Globalisation of Democracy*. Farrar, Straus & Giroux, Macmillan.
Fund For Peace. (2022). *State Fragility Index. P1: Political Indicators*. Fund for Peace. https://fragilestatesindex.org/indicators/p1. (Accessed on 10 February 2023).
Gilens, M., & Page, B. I. (2014). Testing Theories of American Politics: Elites, Interest Groups, and Average Citizens. *Perspectives on Politics, 12*(3).
Gillespie, A. (2004). Good Governance, Corruption and Vote Buying. In International Forums." *The New Zealand Yearbook of International Law*, vol. 1, (103–130)
Gray, C. S. (1996). The Influence of Space Power Upon History. *Comparative Strategy., 15*(4), 293–308.

Gray, C. S., & Sheldon, J. B. (1999). Space Power and the Revolution in Military Affairs: A Glass Half Full? *Airpower Journal., 13*(3), 22–38.

Grosselin, K. (2020). A Culture of Military Spacepower. *Air & Space Power Journal., 34*(1), 75–86.

Gupta, V. (2016). Indian Administrative Service and Crony Capitalism. In N. Khatri, & A. K. Ojha (Eds), *Crony Capitalism in India. Establishing Robust Counteractive Institutional Frameworks.* Palgrave.

Haber, S. (2002). Introduction: The Political Economy of Crony Capitalism. In S. Haber (Ed.), *Crony Capitalism and Economic Growth in Latin America. Theory and Evidence.* Hoover Institution Press.

Hanson, J. K. (2019). Laying the Foundations of the State: Identifying the Constituent Factors of State Building. Paper prepared for delivery at "State Capacity. Concepts-Causes-Consequences." *Duke Workshops in the Frontiers of Political Science.* April 12–13. Duke University.

Hanson, J. K., & Sigman, R. (2021). Leviathan's Latent Dimensions: Measuring State Capacity for Comparative Political Research. *The Journal of Politics*, 83(4).

Harding, R. C. (2012). *Space Policy in Developing Countries: The Search for Security and Development on the Final Frontier*. Taylor & Francis Group.

Harter, M. E. (2006). Ten Propositions Regarding Space Power: The Dawn of a Space Force. *Air & Space Power Journal, 20*(2).

Hays, P. L. (2002). What is Spacepower and Does it Constitute a Revolution in Military Affairs? *Journal of Military and Strategic Studies, 5*(1), 1–37.

Hays, P. L. (2015). Spacepower Theory. In K. U. Schrogl, C. Giannopapa, P. L. Hays, J. Robinson, D. Moura (Eds). *Handbook of Space Security—Policy, Applications and Programs* (1st ed). Springer.

Hays, P. L. (2020). Spacepower Theory and Organizational Structures. In K. U. Schrogl, M. Adriaensen, C. Giannopapa, P. L. Hays, J. Robinson, N. Antoni (Eds). *Handbook of Space Security—Policy, Applications and Programs* (2nd ed.). Springer.

Hays, P. L., Smith, J. M., Van Tassel, A. R., & Walsh, G. M. (2000). *Spacepower for a New Millennium: Space and U.S. National Security*. McGraw-Hill.

Hellman, J. S., & Kaufmann, D. (2001). Confronting the Challenge of State Capture in Transition Economies. *Finance & Development, 38*(3).

Hellmann, J. S., & Kaufmann, D. (2000). *Seize the State, Seize the Day: State Capture, Corruption and Influence in Transition* (Working Paper No. 2444). World Bank Policy Research.

Heper, M. (1985). The State and Public Bureaucracies: A Comparative and Historical Perspective. *Comparative Studies in Society and History* (Vol. 27).

Hitchens, T., & Johnson-Freese, J. (2016). *Toward a New National Security Space Strategy.* Atlantic Council. https://www.atlanticcouncil.org/wp-content/uploads/2015/08/AC_StrategyPapers_No5_Space_WEB1.pdf (Accessed on 2 January 2023).

Hobbes, T. (2017 [1651]). *Leviathan.* Edited by Noel Malcolm. Oxford University Press.

Hoffman, D. E. (2002). *The Oligarchs: Wealth and Power in the New Russia.* Public Affairs.

Hough, J. F. (1977). *The Soviet Union and Social Science Theory.* Harvard University Press.

Hough, J. F. (1997). *Democratization and Revolution in the USSR, 1985–1991.* The Brookings Institution.

Hull, M. M. (2012). Far from Nuremberg. The United States, War Crimes, and the International Criminal Court. *Inter Agency Journal, 3*(2), Special Edition.

Huntington, S. P. (1957). *The Soldier and the State: The Theory and Politics of Civil-Military Relations*. Belknap.

Huntington, S. P. (1968). *Political Order in Changing Societies.* Yale University Press.

Hyatt, J. L., Paul, L. L., Rampino, M. A., Ricchi, R. R., & Schwarz, J. H. (1995). *Space Power 2010.* Air University Press.

Igan, D., & Lambert, T. (2019). *Bank Lobbying: Regulatory Capture and Beyond* (IMF Working Paper [WP/19/171]). International Monetary Fund.

Jacobs, L. R., & Page, B. I. (2005). Who Influences US Foreign Policy? *American Political Science Review., 99*(1), 107–123.

Janowitz, M. (1960). *The Professional Soldier: A Social and Political Portrait*. Free Press.

Johnson, D. J., Pace, S., & Gabbard, B. (1998). *Space: Emerging Options for National Power*. RAND Corporation.

Jusell, J. J. (1998). *Space Power Theory: A Rising Star*. Air University Press.

Kagawa-Fox, M. (2009). Japan's Whaling Triangle—The Power Behind the Whaling Policy. *Japanese Studies., 29*(3), 401–414.

Karlan, P. (1999). Politics by Other Means. *Virginia Law Review, 85*(8), 1697–1724.

Kaufmann, D., Kraay, A., & Mastruzzi, M. (2010). *The Worldwide Governance Indicators: Methodology and Analytical Issues* (World Bank Policy Research Working Paper No. 5430).

Kelley, J. (2007). Who Keeps International Commitments and Why? *American Political Science Review, 101*(3).

Kepron, M., Hitchens, T., and Katz-Hyman, M. (2011). Preserving Freedom of Action in Space: Realizing the Potential and Limits of U.S. Spacepower. In C. D. Lutes, & P. L. Hays (Eds). *Toward a Theory of Spacepower—Selected Essays*. Defense University.

Khatri, N., & Ojha, A. K. (2016). Indian Economic Philosophy and Crony Capitalism. In N. Khatri, & A. K. Ojha (Eds), *Crony Capitalism in India. Establishing Robust Counteractive Institutional Frameworks*. Palgrave.

Khatri, N., & Ojha, A. K. (Eds). (2016). *Crony Capitalism in India. Establishing Robust Counteractive Institutional Frameworks*. Palgrave.

Klein, J. (2006). *Space Warfare: Strategy, Principles, and Policy*. Routledge.

Klein, J. (2019). *Understanding Space Strategy: The Art of War in Space*. Routledge.

Kryshtanovskaya, O., & White, S. (2003). Putin's Militocracy. *Post-Soviet Affairs, 19*(4).

Kumar, R. (2016). Role of Business in India's Foreign Policy. *India Review, 15*(1), 98–111.

Kuziemko, I., & Werker, E. (2006). How Much Is a Seat on the Security Council Worth? Foreign Aid and Bribery at the United Nations. *Journal of Political Economy, 114*(5), 905–930.

Lambakis, S (2001). *On the Edge of Earth: The Future of American Space Power*. University Press of Kentucky.

Larned, R. E. (1994). *Air and Space Doctrine Symposium*. Air University Press.

Lee, J. G. (1994). *Counterspace Operations for Information Dominance*. Air University Press.

Lieberthal, K., & Oksenberg, M. (1988). *Policy Making in China: Leaders, Structures, and Processes*. Princeton University Press.

Linblom, C. E. (1977). *Politics and Markets*. Basic Books.

Linz, J. J., & Stepan, A. C. (1996). Toward Consolidated Democracies. *Journal of Democracy, 7*(2).

Lockwood, N. J. (2013). International Vote Buying. *Harvard International Law Journal, 54*(1), 97–156.

Lupton, D. E. (1988). *On Space Warfare: A Space Power Doctrine*. Air University Press.

Lutes, C. D. (2008). Spacepower in the 21st Century. *Joint Force Quarterly., 49*, 66–72.

Lutes, C. D., & Hays, P.L. (Eds). (2011). *Toward a Theory of Spacepower: Selected Essays*. National Defense University Press.

Mann, M. (1984a). The Autonomous Power of the State: Its Origins, Mechanisms and Results. *European Journal of Sociology, 25*, 185–213.

Mann, M. (1986–2102). *The Sources of Social Power* (4 volumes). Cambridge University Press.

Mann, M. (1993). *The Sources of Social Power. Vol. II. The rise of classes and nation-states, 1760–1914*. Cambridge University Press.

Mann, M. (2013). *The Sources of Social Power. Volume 4: Globalizations, 1945–2011*. Cambridge University Press.

Martindale, M., & Deptula D. A. (2018). *Organizing Spacepower: Conditions for Creating a US Space Force* (Mitchell Institute Policy Papers 16). Mitchell Institute.

Mazumdar, S. (2008). *Crony Capitalism and India: Before and After Liberalization* (Working Paper No. 2008/04). Institute for Studies in Industrial Development.

Migdal, J. S. (1988). *Strong Societies and Weak States: State-Society Relations and State Capabilities in the Third World*. Princeton University Press.

Mills, C. W. (1956). *The Power Elite*. Oxford University Press.

Moltz, J. C. (2019). The Changing Dynamics of Twenty-First-Century Space Power. *Journal of Strategic Security, 12*(1), 15–43.

Morikawa, J. (2009). *Whaling in Japan: Power, Politics and Diplomacy*. Hurst & Company.

Mulgan, A. G. (2000). *The Politics of Agriculture in Japan*. Routledge.

Nettl, J. P. (1968). The State as a Conceptual Variable. *World Politics., 20*(4), 559–592.

Newberry, R. D. (2003). Latin American Countries with Space Programs: Colleagues or Competitors? *Air & Space Power Journal., 17*(3), 39–48.

Nooruddin, I., & Payton, A. L. (2010). Dynamics of Influence in International Politics: The ICC, BIAs, and Economic Sanctions. *Journal of Peace Research., 47*(6), 711–721.

Nye, J. S., Jr. (2004). *Soft Power. The Means to Success in World Politics*. Public Affairs Books.

Oberg, J. E. (1999). *Space Power Theory*. US Air Force Academy.

Office of the US Trade Representative (2000). *Monitoring and Enforcing Trade Laws and Agreements*. Fact Sheet, Executive Office of the President. https://1997-2001.state.gov/issues/economic/fsustr_000501_tradelaws.html (Accessed on 2 January 2023).

Olson, M. (1965). *The Logic of Collective Action: Public Goods and the Theory of Groups*. Harvard University Press.

Orttung, R. W. (2006). The Role of Business in Russian Foreign and Security Relations. In A. Wenger, J. Perovic, & R. Orttung (Eds). *Russian Business Power. The Role of Russian Business in Foreign and Security Relations*. Routledge.

Orttung, R. W. (2004). Business and Politics in the Russian Regions. *Problems of Post-Communism., 51*(2), 48–60.

Paikowsky, D. (2017). *The Power of the Space Club*. Cambridge University Press.

Palmer, G., Wohlander, S. B., & Morgan, T. C. (2002). Give or Take: Foreign Aid and Foreign Policy Substitutability. *Journal of Peace Research, 39*(1), 5–26.

Paracer, S., & Ahmadjian, V. (2000). *Symbiosis: An Introduction to Biological Associations*. Oxford University Press.

Pfaltzgraff, R. L. (2011) International Relations Theory and Spacepower. In C. D. Lutes, & P. L. Hays (Eds). *Toward a Theory of Spacepower: Selected Essays*. National Defense University Press.

Rana, K. S. (2017). Serving the Private Sector: India's Economic Diplomacy. In N. Bayne, & S. Woolcock (Eds). *The New Economic Diplomacy: Decision-Making and Negotiation in International Economic Relations*. Routledge.

Remington, T. (2008). Patronage and the Party of Power: President-Parliament Relations Under Vladimir Putin. *Europe-Asia Studies., 60*(6), 959–987.

Roth, K. (2003). *Letter to U.S. Secretary of State Colin Powell on Exemption Agreement Waivers for All ICC States Parties* (9 December). https://www.hrw.org/node/235872/printable/print (Accessed on 2 January 2023).

Rueschemeyer, D., & Evans, P. (1985). The State and Economic Transformation: Toward an Analysis of the Conditions Underlying Effective Intervention. In P. Evans, S. Rueschemeyer, T. Skocpol (Eds). *Bringing the State Back In*. Cambridge University Press.

Sartori, G. (1970). Concept Misformation in Comparative Politics. *The American Political Science Review, 64*(4), 1033–1053.

Savage, C. (2023). Pentagon Blocks Sharing Evidence of Possible Russian War Crimes With Hague Court. *The New York Times* (March 8).

Schattschneider, E. E. (1960). *The Semisovereign People*. The Dryden Press.

Shlapentokh, V. (1996). Early Feudalism: The Best Parallel for Contemporary Russia. *Europe–Asia Studies, 48*(3).

Shlapentokh, V. (2003). Hobbes and Locke at Odds in Putin's Russia. *Europe–Asia Studies, 55*(7).

Shlapentokh, V. (2007). *Contemporary Russia as a Feudal Society*. Palgrave–Macmillan.

Skocpol, T. (1985). Bringing the State Back in: Strategies of Analysis in Current Research. In P. Evans, D. Rueschemeyer, & T. Skocpol (Eds). *Bringing the State Back In*. Cambridge University Press.

Smith, M. V. (2001). *Ten Propositions Regarding Spacepower*. Air University Press.

Smith, M. V. (2016). *Spacepower and the Strategist*. Naval Institute Press.
Stigler, G. (1971). The Theory of Economic Regulation. *Bell Journal of Economics and Management Science., 2*, 3–21.
Stiglitz, J. E. (2002). *Globalization and its Discontents*. W. W. Norton & Co.
Stone, R. W., Wang, Y., & Yu, S. (2022). Chinese Power and the State-Owned Enterprise. *International Organization., 76*(1), 229–325.
Strand J. R., & Tuman, J. P. (2012). Foreign Aid and Voting Behavior in an International Organization: The Case of Japan and the International Whaling Commission. *Foreign Policy Analysis, 8*(4).
Strüver, G. (2016). What Friends Are Made of: Bilateral Linkages and Domestic Drivers of Foreign Policy Alignment with China. *Foreign Policy Analysis, 12*(2), 170–191.
Sullivan, P. L., Tessman, B. F., & Li, X. (2011). US Military Aid and Recipient State Cooperation. *Foreign Policy Analysis, 7*, 275–294.
Surindar, P., & Ahmadjian, V. (2000). Symbiosis: *An Introduction to Biological Associations*. Oxford University Press.
Teorell, J., et al. (2020). *The Quality of Government Standard Dataset, Version Jan 20*. University of Gothenburg.
Tilly, C. (Ed.). (1975). *The Formation of National States in Western Europe*. Princeton University Press.
Tortora, J. J. (2015). "European Autonomy in Space: The Technological Dependence." In C. Al-Ekabi (Ed.). *European Autonomy in Space*. Springer.
Townsend, B. (2019). Space Power and the Foundations of an Independent Space Force. *Air & Space Power Journal, 33*(4), 11–24.
Transparency International. (2022). Corruption Perceptions Index. *Transparency International*. https://www.transparency.org/en/cpi/2022. (Accessed on 10 February 2023).
USAF. (1998). *Air Force Doctrine Document 2-2*. Air Force Doctrine Center. United States Air Force.
USSF. (2020). *Spacepower: Doctrine for Space Forces* (Military Doctrine). Headquarters United States Space Force.
Vaccaro, A. (2022). Measures of State Capacity: So Similar, Yet So Different. *Quality & Quantity*.
Varma, A., Hu, B., & Bloomquist, L. (2016). Family Oligarchies and Crony Capitalism in India. In N. Khatri, A. K. Ojha (Eds). *Crony Capitalism in India. Establishing Robust Counteractive Institutional Frameworks*. Palgrave.
Vreeland, J. R. (2019). Corrupting International Organizations. *Annual Review of Political Science, 22*, 205–222.
Weber, M. (1994). The Profession and Vocation of Politics. In P. Lassman, & R. Speirs (Eds). *Weber: Political Writings*. Cambridge University Press.
Wedel, J. R. (2009). *Shadow Elite. How the Worlds New Power Brokers Undermine Democracy, Government, and the Free Market*. Basic Books.
Wedel, J. R. (2014). *Unaccountable. How the Establishment Corrupted Our Finances, Freedom and Politics and Created an Outsider Class*. Pegasus Books.
Winters, J. A. (2011). *Oligarchy*. Cambridge University Press.
Winters, J. A., & Page, B. I (2009). Oligarchy in the United States? *Perspectives on Politics, 7*(4).
Woo, B., & Chung, E. (2018). Aid for Vote? United Nations General Assembly Voting and American Aid Allocation. *Political Studies., 66*(4), 1002–1026.
Wouters, J., & Hansen, R. (2015). Strategic Autonomy in EU Space Policy: A Conceptual and Practical Exploration. In C. Al-Ekabi (Ed.). *European Autonomy in Space*. Springer.
Yackee, J. W., & Yackee, S. W. (2006). A Bias Towards Business? Assessing Interest Group Influence on the U.S. Bureaucracy. *The Journal of Politics., 68*(1), 128–139.
Zeng, K. (2004). *Trade Threats. Bargaining, Retaliation, and American Coercive Diplomacy*. University of Michigan Press.

Chapter 3
Measuring Space Actors: A Methodological Framework

This chapter offers a methodology to operationalise the conceptual framework described in Chap. 2. It provides a detailed description of how spacepower can be disaggregated into its two constituent dimensions of capacity and autonomy, and how these can be further broken down into four subdimensions: hard capacity, soft capacity, hard (or technical) autonomy and soft (or political) autonomy. For each of these subdimensions, the chapter features an in-depth description of the constituent indicators and provides a detailed explanation of the underlying measurement model.

3.1 Overview of the Methodological Framework and Scoring System

In Chap. 2, we provided evidence for the adoption of a multi-dimensional understanding of *spacepower* and, unlike advocates of a rigid division in tier groups, we proposed a more complex, ideal–typical taxonomy of space actors visualised as distributed in a non-hierarchical space. Our spacepower matrix accomplishes this goal. The matrix describes how the concept of spacepower can be operationalised and why only some states may claim *space power* status. Indeed, the matrix is based on the combination of two analytical dimensions—autonomy and capacity—and four subdimensions in which the concept of spacepower can be articulated and by which it can be measured. That said, this conceptualisation requires a way to empirically map space actors within the matrix. In this chapter we develop a measurement model that can be used to empirically measure and compare the status of space actors within the spacepower matrix.

The process leading to the translation of multi-dimensional concepts into composite or synthetic measures is generally based on four major stages. The first stage consists in providing a definition of the core concept. As we have contended, space power refers to the status of those actors having the means to deploy, operate,

M. Aliberti et al., *Power, State and Space*, Studies in Space Policy 35,
https://doi.org/10.1007/978-3-031-32871-8_3

and benefit from any space-related capabilities to support the achievement of national goals that they have autonomously determined.

The second stage involves the identification of the various dimensions in which the core can be articulated. In our theoretical framework, these correspond first and foremost to capacity and autonomy.

The third stage consists in selecting the indicators relevant to the observation of the various dimensions of interest. According to a widely accepted definition, indicators are "indirect empirical representations used to define or refer to concepts when no direct measurement is possible" (Payne & Payne, 2004). Indicators should point to directly observable characteristics and be unambiguously linked to the dimensions of interest. As Payne and Payne (2004: 118) put it, indicators "must properly reflect the essential nature of the core concept".

The fourth stage results in combining, when possible, the selected indicators into a synthetical measure: that is to say, aggregating multiple variables into one composite indicator or index. Different aggregation methods clearly produce different outcomes. Moreover, each aggregation method implies a set of theoretical assumptions and should be picked carefully: for instance, averaging two indicators produces a compensation effect, as any increase in one variable is meant to balance out an equal decrease in the other one.

Whereas Chap. 2 focused on the first two stages, this chapter expands on the remaining two. The methodology adopted within this book is straightforward. As it has been already pointed out, the concept of space power covers two fundamental dimensions: capacity and autonomy. Each dimension includes two sub-dimensions. Specifically, capacity has been broken down in hard and soft capacity, and autonomy in technical and political autonomy. Each sub-component has been consequently articulated in a number of areas. Each area encompasses several entries. Eventually, each entry has been attributed a score.

The scoring system involves a cascading process, with scores of entries initially flowing into areas and ultimately leading to a composite "space power index". The compilation of the index is articulated into six major phases and is illustrated in Fig. 3.1.

First, each indicator (e.g., number of satellites) has been given a score on a scale running from 1 to 4, with 1 corresponding to the lowest level of capacity or autonomy and 4 corresponding to the highest level of capacity or autonomy. Second, scores of each indicator have been averaged for the area to which they belong (e.g., telecommunications), resulting in a single value ranging from 1 to 4. Third, scores of each area have been averaged for the macro-area to which they belong (e.g., application satellites), resulting in a single value ranging from 1 to 4. Fourth, scores of each macro-area have been averaged for the sub-dimension to which they belong (e.g., hard capacity), resulting in a single value ranging from 1 to 4. Fifth, scores of each sub-dimension have been averaged for the dimension to which they belong (e.g., capacity), resulting in a single value ranging from 1 to 4. Sixth, scores of both autonomy and capacity have been jointly plotted to form a "Spacepower matrix," as well as combined into the so-called "Spacepower index."

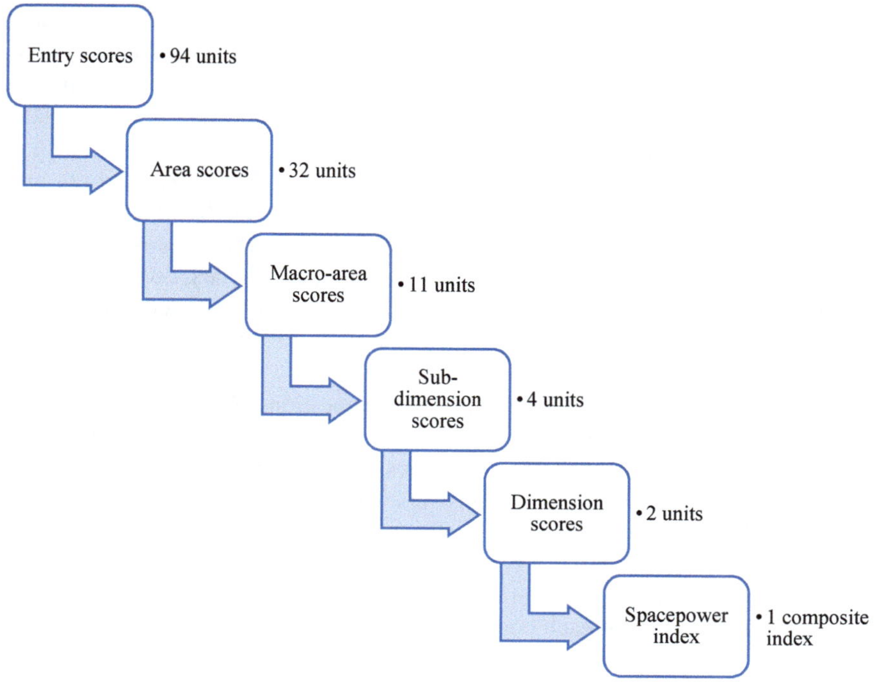

Fig. 3.1 Compilation of the "Space power index"

The spacepower index relies on both quantitative and qualitative data. On the one hand, quantitative data are relevant to the observation of hard capacity and technical autonomy; on the other, qualitative data allow to operationalise soft capacity and political autonomy.

Quantitative data have been retrieved from a variety of specialised sources and combined with the database of the European Space Policy Institute (ESPI) in Vienna. Importantly, the data extend over a span of time of 10 years, from 2011 to 2020. This dataset has been built by collecting open-source data on rocket launches for the period under consideration, taking into account the launchers, launch sites, launching countries, and the launch outcome on the launcher side. For every launch, the dataset lists payloads, their manufacturer and operator, their nationality, and their function in categories. Data not directly connected with launches and satellite systems have been gathered thanks to reports of organisations and think tanks such as the Secure World Foundation, the Center for Strategic International Studies, the Union of Concerned Scientists or the Science and Technology Policy Institute, as well as integrated with other open-source data. In order to make variables comparable with one another, scores for each dimension have been converted to a four-point scale ranging from 1 to 4.

Qualitative data was collected thanks to the elaboration and administration of an ad hoc survey, the "measuring spacepower" survey, administered to selected experts on the space activities of the space actors considered in this work. Expert-based evaluations have increased the validity of the assessment presented in this study, allowing to provide insights into the "soft" and "political" dimensions, compensating for the lack of extensive or high-level data on indicators concerning space-related decision-making processes.

The "measuring spacepower" survey was launched in April 2021 and ran until October 2021. All responses refer to the period ranging from 2010 to 2020 (before the outbreak of the COVID-19 pandemic). The questionnaire has been accessible by invitation only, with invitation emails being sent to prominent scholars, decision-makers, and analysts in the field of space studies who possess demonstrated geographical expertise relevant to the conduct of the comparative assessment. All answers have been anonymised.

The survey is composed of 50 multiple choice questions. Respondents have been asked to provide answers strictly for their country of expertise. Answers to the survey have involved the attribution of scores on a four-point scale running from 1 to 4. Overall, about 50 anonymised answers were collected, with priority given to keeping a balanced number of answers collected for each space actor included in the analysis. In an effort to ensure a balanced evaluation and try to avoid any normative and/or political bias, we ensure that respondents for each country were based both inside and outside of each country's national borders. In the peculiar case of Europe, selected respondents were based in six different European countries (Austria, Belgium, France, Germany, Italy, and the United Kingdom). The complete list of the questions featured in the "Measuring space power" survey has been included in Appendix C.

3.2 Measuring Capacity

As presented in Chap. 2, the first of the two dimensions in which the concept of space power has been broken down relates to the sphere of capacity. In the context of space activities, capacity can be broadly understood as the state's ability to *implement* space-related strategies in order to achieve its economic, political, or societal goals.

Capacity comprises both a hard and a soft sub-dimension. To capture shifts in the overall capacity distribution, a set of two sub-indices has been deployed:

- The *hard capacity sub-index*, based on 37 quantitative indicators, which aims to assess the material assets and abilities allowing an actor to operate in, through, and from space across the full spectrum of existing space activities;
- The *soft capacity sub-index*, based on 26 qualitative indicators, which aims to evaluate the state's ability to effectively utilise and integrate assets and expertise in national policies, infrastructure, and activities.

Both sub-indices comprise various areas, each encompassing several entries. As shown above, a score ranging from 1 ("lowest capacity") to 4 ("highest capacity") has been attributed to each entry. Entries score have been consequently averaged for

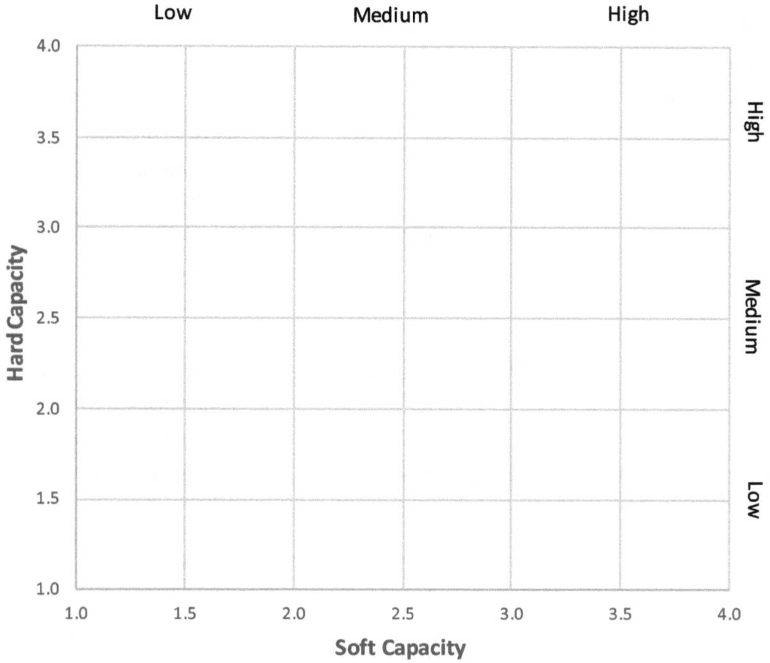

Fig. 3.2 Capacity sub-matrix overview

the corresponding area, ultimately leading to the compilation of a synthetic indicator for the "hard capacity" sub-dimension and a synthetic indicator for the "soft capacity" sub-dimension.

The values obtained from the measurement of the two capacity indices will be subsequently transposed into a sub-matrix that graphically combines a country's hard and soft capacities and presents its position vis-à-vis other countries (Fig. 3.2).

Eventually, scores for the two capacity sub-indices have been averaged and combined into a single capacity index. This total capacity value will be used to plot the main matrix of space power.

3.2.1 Hard Capacity

What we call hard capacity stretches across four prominent macro-areas: "applications satellites," "science and exploration," "space safety and security," and "enabling and support." Each macro-area, in turn, covers a number of areas, for a total in 12 areas of interest. Each area includes a number of indicators. Table 3.1 includes a comprehensive list of the indicators upon which the measurement model is based. As for other sub-dimensions featured in the book, an in-depth explanation of each indicator is provided in the dedicated sections of the chapter in hand.

Table 3.1 Measuring hard capacity: macro-areas, areas, and entries

Macro-area	Area	Entry
Application satellites	Telecommunications	1: Number of satellites
		2: Mission diversity
		3: Performance (technological assortment)
	Remote sensing	4: Number of satellites
		5: Mission diversity
		6: Performance (variety of sensor types)
	Positioning, navigation timing	7: Coverage
		8: Augmentation (space-based)
		9: Performance (accuracy)
Science and exploration	Space and Earth science	10. Number of space science missions
		11: Number of Earth science missions
	Human spaceflight	12: Astronauts
		13: Cargo vehicles/programs
		14: Crew vehicles/programs
		15: Space stations infrastructures
		16: Extra-vehicular activities (EVAs)
	Robotic exploration	17: Number of exploration missions
		18: Number of operational missions
		19: Diversity of destination
		20: Mission architecture variety
Space safety and security	Space situational awareness	21: Space surveillance and tracking
		22: Near earth objects monitoring
		23: Space weather
	Counter-space	24: R&D of counter-space capabilities
		25: Testing of counter-space capabilities
		26: Deployment of counter-space capabilities
Enabling and support	Space transportation	27: Launch cadence
		28: Launchers' variety

(continued)

Table 3.1 (continued)

Macro-area	Area	Entry
		29: Launchers' performance
		30: Launchers' reliability
	Ground operations	31: TT&C infrastructure
		32: Launch facilities
	Space operations	33: Active-debris removal capabilities
		34: Life extension services
		35: In-space deployment capabilities
	Tech demonstrator	36: Number of tech demonstrator missions
		37: Nature of tech demonstrators

3.2.1.1 Application Satellites

Space applications have become an integral part of modern societies to such an extent that a world suddenly deprived of its navigation, telecommunications or Earth observation satellites would most certainly plunge into chaos. On a daily basis, countless governmental, military, commercial and individual users rely on space-based applications for a myriad of services. As Pelton et al. (2013) summarise it:

> Services such as worldwide news, satellite entertainment channels, coverage of sporting events, communications to ships at sea or aircraft in the skies, and many more frequently depend on satellites. Farmers now rely on satellites to irrigate their crops, add the right amount of fertilizer, or detect a crop disease. Fishing fleets use satellites to know where to fish. Energy and resource companies employ satellite imaging to know where to dig or drill. Efforts to combat global warming, preserve the Ozone layer that is essential to life on Earth, and other activities to sustain the biodiversity of plant and animal life on our planet all depend on applications satellites. Responding to major disasters routinely involves analysis of satellite imagery and mobile satellite communications.

All these applications are enabled by application satellite systems, which can be of different nature and serve the most diverse purposes. Based on commonly used categorisations, the macro-area "application satellites" can be segmented in three main areas, namely telecommunications, remote sensing and positioning, navigation and timing. A description of these three areas is provided hereby.

Telecommunications Systems

Telecommunications satellites are the most known and mature type of application satellites. The first practical application of space-based technology in the 1960s was in the field of telecommunications. These satellites enable global telecommunications systems by receiving and relaying signals with voice, video and data from and to one or more locations on the ground. While Earth-based alternatives to satellite communications are generally available, their use is often critical during natural disasters and emergencies when other land-based delivery systems are down (Labrador, 2023). In addition, as stressed by the United Nations Office for Outer Space Affairs, "space-based technology can often reduce infrastructure requirements and offer more cost-effective service delivery options. For instance, instead of constructing a series of transmission and relay towers to broadcast television programmes to far-to-reach places, one satellite dish could be provided to a remote community to pick up broadcast signals sent from a satellite" (UNOOSA, 2023a). Telecommunication satellites can be of many different types, "with designs varying according to their purpose. They use different orbits, different frequencies and they transmit very different types of signals using a variety of power levels" (ESA, 2023a). The telecommunications area has been measured through the following indicators:

Indicator 1: Number of Telecommunication Satellites. This indicator weighs the number of telecom-related satellites launched in the period 2011–2020 and operated by the country. When the number of satellites reaches 10 per year or more (at least 100 in the period considered), the country and its national actors showcase a significant activity in satellite communication, demonstrating that satcom is a relevant part of the national telecommunications infrastructure. Scores for this entry have been attributed as follows:

4 = more than 100 satellites (not considering mega-constellations).
3 = between 21 and 100 satellites (not considering mega-constellations).
2 = between 2 and 20 satellites.
1 = between 0 and 1 satellite.

Indicator 2: Mission Diversity. Telecommunications satellites can be classified in different categories. Classification could be done based on whether satellites are providing fixed or mobile satellite services, or based on their mission. Telecommunications satellites can be typically divided in (a) broadcasting and general communication; (b) automatic information system (AIS); (c) data relay. Scores for this entry have been attributed as follows:

4 = three missions (or more).
3 = two missions.
2 = one mission.
1 = no satcom launched.

Indicator 3: Performance—Technological Range. Telecommunications satellites can have different architectures and involve different technologies. Recent technological developments include (a) LEO broadband constellations; (b) high-throughput satellites; (c) space-to-space data relay; (d) IoT from space; (e) military satellites, which usually implement advanced technologies and deliver specific advanced missions, given their security requirements. This indicator aims to assess this range by rewarding countries that try to push satcom technological innovation and deploy a vast assortment of technologies, pursuing one of the innovative satellite architectures/ systems mentioned above. Scores for this entry have been attributed as follows:

4 = three or more types of innovative tech.
3 = two types of innovative tech.
2 = one type of innovative tech.
1 = no innovation (just regular FSS/MSS Satcom).

Remote Sensing Systems

Remote sensing is the process of detecting, monitoring and acquiring information about the physical characteristics of an area by measuring its reflected or emitted radiation without direct contact with the area typically through the use of satellites or aircraft. Remote sensing spacecraft work as an enlarged human eye, enabling planetary scale measurements and consistent replications of these measurements over the years (NASA, 2014). Remote sensing satellites can be used for a variety of military and civil purposes, including meteorology, early warning; environmental and resource monitoring, cartography and urban planning, research in geosciences, and disaster management (Madry, 2013). As for telecommunications, to showcase a significant activity in remote sensing a certain number of satellites is indispensable. Furthermore, due to the mission variety, a wide range of sensors is required to gather imaging data and extract different types of information that can serve a very wide range of applications and industries (EUSPA, 2021a). The variety of sensors type is therefore an important indicator. Within our measurement model, the remote sensing area has been assessed through the following indicators:

Indicator 4: Number of Remote Sensing Satellites. This indicator measures the number of remote sensing satellites launched in the period 2011–2020 and operated by the considered actor. When the number of satellites reaches 10 per year or more (at least 100 in the period considered), we consider that the country and its national entities showcase significant activity in remote sensing. Scores for this entry have been attributed as follows:

4 = more than 100 satellites.
3 = between 20 and 100 satellites.
2 = between 2 and 20 satellites.
1 = 0 or 1 satellite.

Indicator 5: Mission Diversity. Remote Sensing satellites can be classified in different categories. Based on publicly available data and common classifications, categories considered are (a) Earth Observation; (b) Meteorology; (c) Early Warning; (d) Signal Intelligence; (e) Space Situational Awareness. Scores for this entry have been attributed as follows:

4 = three missions or more.
3 = two missions.

2 = one mission.
1 = no remote sensing satellites.

Indicator 6: Performance (Variety of Sensor Types). The variety and different level of performance offered by remote sensing satellites is vast and rapidly developing. While the information on resolution available to countries is not always accessible, due to military or proprietary restrictions, or cumbersome to compare in a simple way, a proxy to highlight performance can be the variety of sensors that a country has access to. The wider the spectrum of sensors deployed, the more applications a nation can have thanks to the satellites. This indicator is assessed considering optical and radar as a baseline for remote sensing operations. Adding meteorology-related space sensors to the baseline helps achieve an optimal operational capability. To this optimal, one country can add additional advanced sensors such as hyperspectral cameras or RF monitoring sensors, to obtain the highest score. Scores for this entry have been attributed as follows:

4 = Optimal + additional sensors (Hyperspectral, RF monitoring, etc.)
3 = Operational optimal (Radar + Optical + Meteorology).
2 = Operational baseline (Radar + Optical).
1 = Below baseline.

3.2.1.2 Positioning, Navigation, Timing Systems

Satellite navigation is the "youngest" of the major types of satellite applications (Pelton & Camacho-Lara, 2013). Nonetheless, it has rapidly become an integral and critical tool for the functioning of modern societies, with countless of governmental, military, commercial, and even individual users that everyday rely on navigation systems for a variety of purposes. As pointed out by the US Department of Transportation, what is commonly referred to as a navigation refers to a "combination of three distinct, constituent capabilities:

- Positioning, or the ability to accurately and precisely determine one's location and orientation two-dimensionally (or three-dimensionally when required) referenced to a standard geodetic system (such as World Geodetic System 1984, or WGS84);
- Navigation, or the ability to determine current and desired position (relative or absolute) and apply corrections to course, orientation, and speed to attain a desired position anywhere around the world, from sub-surface to surface and from surface to space; and
- Timing, or the ability to acquire and maintain accurate and precise time from a standard (Coordinated Universal Time, or UTC), anywhere in the world and within user-defined timeliness parameters. Timing also includes time transfer" (US Department of Transportation, 2023).

The provision of positioning, navigation and timing (PNT) services is enabled by different types of space-based systems known as GNSS (Global Navigation Satellite System) or RNSS (Regional Navigation Satellite System). The term GNSS designates a constellation of satellites providing signals from space that transmit positioning and timing data to GNSS receivers, which then use this data to determine location (EUSPA, 2021b). By definition, a GNSS like the American GPS or the Russian GLONASS provide global coverage, while RNSS like the Japanese QZSS or the Indian NAVIC a regional one.

Within our measurement model, the PNT area has been assessed through the following indicators:

Indicator 7: System Coverage. Navigation Satellite Systems can be regional or global in nature. Hardware and infrastructure required for a regional system are more limited compared to that of a global system. This indicator assesses whether the considered actors fully deployed a navigation satellite system by 2020, as well as on whether they established a regional or global navigation satellite system. Scores for this entry have been attributed as follows:

4 = fully deployed global system.
3 = regional coverage (deployed regional system, or global system under deployment).
2 = regional system under deployment.
1 = no GNSS system.

Indicator 8: Augmentation (Space-Based). In addition to navigation satellite systems, countries use augmentation systems to improve accuracy of the PNT services provided by satellite. Few countries in the world have developed satellite-based augmentation systems (SBAS), others (as New Zealand or Australia) have pledged to develop it. The deployment of SBAS systems requires further infrastructure and it is an enabler for more accurate operations, in particular in the aviation industry. This indicator assesses whether the considered actors have developed and deployed a SBAS. Scores for this entry have been more specifically attributed as follows:

4 = SBAS system deployed.
3 = SBAS system under deployment.
2 = SBAS system under development.
1 = No system.

Indicator 9: Performance (Accuracy). GNSS have many different elements that can be compared. The most immediate one, and a proxy of their potential for applications, is the accuracy to the metre of the systems. GNSS have usually a twofold accuracy: one for its publicly available service and one encrypted service, usually delivered to authorised military entities only. This indicator scores whether one country's navigation satellite system has achieved sub-metre accuracy. Scores for this entry have been attributed as follows:

4 = sub-metre accuracy.
3 = 10 m and lower, up to one metre.
2 = 10 m or higher.

1 = no system.

3.2.1.3 Science and Exploration

Science and exploration form an integral component of *spacepower* and of major space actors' quest to assert themselves as *space powers*. Since the beginning of the space age, science and exploration have been a catalyst for innovation, inspiration, human progress and economic development (ESA, 2023c). The rationales driving science and exploration activities are manifold and include "pragmatic rationales, which involve economic benefits, contributions to national security, contributions to national stature and international relations, inspiration for students and citizens to further their science and engineering education, and contributions to science; and aspirational rationales, which involve the eventual survival of the human species (through off-Earth settlement) and shared human destiny and the aspiration to explore" (UNCOPUOS, 2018).

In order for a country to fully exploit the potential of space and develop appropriate capabilities, undertaking activities within this macro-area is of critical importance, as scientific research typically leads to the necessary development of new technologies, skills and know-how. As a matter of fact, there is a mutually beneficial interplay between the two dimensions, well summarised in the ditty: science leads technology and technology promotes science.

The type of activities covered in this macro-area are as diverse as closely intertwined. It includes solar and space physics; human and robotic exploration of planets, moons, and asteroids in the solar system; study of the origin, evolution, and current state of objects and phenomena in the universe beyond the solar system; research on non-living and living materials, including humans, under microgravity conditions; as well as the study of the Earth from space (Logsdon, 2023). For the sake of simplicity, this macro-area has been segmented in three main areas, namely: science (which covers both space and Earth sciences), human spaceflight activities, and robotic exploration. A description of these three areas is provided hereby.

Space and Earth Science

Space science, a broad term indicating the application of physical sciences to all things space, is an area playing an important role within the programmes of major space agencies across the world. Whereas the number of missions dedicated to space science is more limited in comparison to the satellites launched for telecommunications or remote sensing, their role is not secondary importance to the build-up of *spacepower*. As ESA (2023c) finely puts it, "space science is a subject that strives to answer the ultimate questions: how did our Earth and our Solar System form and evolve? What is our place in the Universe? Where are we going? Where did life come from, and are we alone? [...] Space science missions provide access to the largest science laboratory we have ever known: our Universe."

Space sciences encompasses many different branches. One the most developed is astronomy, which is the study and measurement of the physical properties of extra-terrestrial objects and phenomena. Astronomy itself encompasses many subfields, such as astrophysics, astrobiology, astrochemistry, cosmology and planetary science.

While space science deals with the exploration and study of space, Earth science is concerned with the study of the Earth itself. Those subjects that deal with the water and air at or above the solid surface of Earth. These include "the study of the water on and within the ground (hydrology), the glaciers and ice caps (glaciology), the oceans (oceanography), the atmosphere and its phenomena (meteorology), and the world's climates (climatology)" (Windley & Albritton, 2021). Space assets, Earth observation satellites in particular, provide fundamental insights into these disciplines and have often become irreplaceable means to study the Earth system. For instance, "information derived from satellite data can contribute to more than half of the 54 ECVs identified by the Global Climate Observing System" (ESA, 2023d).

Given the diversity of missions under consideration and the issues in creating a uniform assessment, within our measurement model, the "space and Earth science" area has been assessed through the following two indicators:

Indicator 10: Number of Astronomy and Space Science Missions. This indicator measures the number of Astronomy and Space Science-related satellites launched in 2011–2020 and operated by the country. Given the technologies and specialised human capital required for such missions, the number of satellites launched is well below the number of those with telecommunications or remote sensing applications. Countries that have launched at least one mission every year score the highest value. Scores for this entry have been more specifically attributed as follows:

4 = launched at least one mission per year (>= 10).
3 = launched at least one mission every other year (>= 5).
2 = launched at least one mission.
1 = no mission launched.

Indicator 11: Number of Earth Science Missions. Number of Earth science-related satellites launched in 2011–2020 and operated by the country. Given the technologies and specialised human capital required for such missions, the number of satellites launched is well below the number of those with telecommunications or remote sensing applications. Countries that have launched at least one mission every year score the highest. Scores for this entry have been attributed as follows:

4 = launched at least one mission per year (>=10).
3 = launched at least one mission every other year (>=5).
2 = launched at least one mission.
1 = no mission launched.

Human Spaceflight

More than 50 years after the ground-breaking landing of Apollo 11 on the surface of the Moon, human spaceflight activities remain important "markers" that identify great power status in the contemporary international system (Cunningham, 2009). Crewed missions indeed represent highly inspiring and internationally visible achievements that strongly shape public perceptions, put the country at the forefront in the international arena and contribute to its image as a major player (Aliberti, 2015).

Human spaceflight programmes, however, continue to be extremely hazardous undertakings that not simply require large amounts of financial expenditure and strong political commitment; they also involve daunting technological feats, inherently fraught with dangers of failure and human loss. It is no coincidence that no human has travelled beyond Low Earth Orbit since the concluding mission of the Apollo programme in December 1972.

While the number of countries that have their astronauts, cosmonauts or taikonauts flown in space has grown over the past decades, the magnitude of human, technical and financial resources that is required remains such that only a few countries have been able to fully master the combination of knowledge, capabilities and infrastructure necessary for sustaining these costly, complex and highly ambitious programmes. As of 2023, the United States, Russia, and China are the only countries with the autonomous capacity to undertake human spaceflight activities.

Within our measurement model, the "human spaceflight" area has been assessed through the following indicators:

Indicator 12: Astronauts. Training astronauts and preparing them for space missions require a significant effort in terms of facilities and workforce involved. The number of astronauts that reached space during the period considered can be an indicator for a country's efforts in human Spaceflight activities. Astronauts' missions are conducted in three (or, more rarely, two) at a time, meaning that a country launching less than two astronauts per year is likely joining a third party's mission rather than conducing one itself, or conducting human spaceflight activities just occasionally. By the end of 2020, only Russia and the USA have been regularly conducting fully crewed space missions. Scores for this entry have been attributed as follows:

4 = two or more astronauts launched per year.
3 = at least one astronaut launched every other year.
2 = at least one astronaut launched.
1 = no astronaut launched in 2011–2020.

Indicator 13: Extra-Vehicular Activities. Conducting Extra-Vehicular Activities (EVAs) requires significant technology, workforce, training, and coordination. EVAs are incredibly hard for astronauts and ground personnel alike, and efforts in doing EVAs are not to be taken for granted. To compare countries based on EVAs activities, the total number of EVAs conducted in 2011–2020, with a consideration for their rhythm (less than one per year, at least one per year, more than one every quarter) can provide an idea of this effort—knowing that this could be biased as only US and Russia (and China) have full capability to conduct EVAs autonomously. Scores for this entry have been attributed as follows:

4 = took part in more than one spacewalk per quarter in the period considered (more than 40 EVAs in 2011–2020).
3 = took part in more than one spacewalk per year in the period considered (more than 10 EVAs in 2011–2020).
2 = took part in at least one spacewalk in the period considered.
1 = no spacewalks by national astronauts/cosmonauts/taikonauts in 2011–2020.

Indicator 14: Cargo Vehicles. Cargo capsules are essential in delivering supplies to space stations or other destinations in orbit. Countries might have know-how due to past missions (e.g., Italy, Europe) without having cargo capsules in active development/use. Countries might have proven hardware and capsules, albeit not making frequent use of it (Japan, with its HTV, for instance). Countries could also make extensive use of these capsules. This indicator tries to address the use of cargo capsules and the ownership of related hardware and know-how. Scores for this entry have been attributed as follows:

4 = frequent use of hardware.
3 = hardware, but limited operations in the period considered.
2 = expertise/hardware.
1 = no expertise.

Indicator 15: Human-Rated Vehicle/Programmes. Like cargo capsules, crew capsules and crew transportation systems are vital for human spaceflight. Some countries might have inherited know-how but have no active programmes pursuing human spaceflight capabilities, while others might be developing crewed capsules but have no spaceflight experience yet. Finally, few countries have developed and are currently and regularly utilizing crewed capsules. Scores for this entry have been attributed as follows:

4 = Active capsules.
3 = active development.
2 = expertise/hardware.
1 = no expertise.

Indicator 16: Space Stations Infrastructures. Building and maintaining space stations require specific expertise, know-how and resources. Some countries can afford to develop space stations on their own, others can decide to focus on making large and varied contributions to international endeavours. Others can decide to specialise on niche applications to deploy connected to a space station (Canada, for instance, is specialised in robotic arms, such as the current CanadArm-2 on board the ISS), to have a small, albeit essential presence in orbit. Scores for this entry have been attributed as follows:

4 = full-fledged space stations.
3 = modules and/or infrastructures.
2 = specialised devices.
1 = no contributions to space stations programmes.

Robotic Exploration

Robotic exploration is an important component of the science and exploration macro-area. Besides being a powerful driver for advancing science and technology, allowing for great discoveries and deepening the understanding of the origins, evolution and destination of our universe, robotic exploration missions serve as precursors to human exploration, contributing to the necessary knowledge that will make subsequent human missions in the solar system safer and more productive (UNCOPUOS, 2018). As also highlighted by the International Space Exploration Coordination Group in its Global Exploration Roadmap, "precursor robotic missions are essential for filling strategic knowledge gaps related to safe and successful human missions and for ensuring maximum return on the investments required for subsequent human exploration. Activities to increase the synergy between human and robotic missions driven by science and human exploration goals remain a high priority" (ISECG, 2013).

As for human spaceflight activities, robotic space exploration requires the development and advancement of key enabling technologies, reliable systems and efficient operation concepts that make such missions particularly demanding in terms of human and financial resources. The number of missions undertaken in this area is therefore more limited compared to applications satellite systems. Not surprisingly, only a handful of players has been so far able to master a combination of the knowledge, capabilities and infrastructure necessary for robotic exploration.

Within our measurement model, the "robotic exploration" area has been assessed through the following indicators:

Indicator 17: Number of Missions Launched in 2011–2020. Number of robotic exploration missions launched in 2011–2020 and operated by the country. Given the technologies and specialised human capital required, the number of missions launched is well below the number of launched satellites with telecommunications or remote sensing applications. Even launching a single mission of this type can require nation-wide efforts, and only a handful of countries have the resources to develop and launch multiple missions in a relatively short amount of time.

Scores for this entry have been attributed as follows:

4 = launched at least one mission every other year (>=5).
3 = launched more than one mission and less than 5 missions.
2 = launched one mission.
1 = no mission launched.

Indicator 18: Number of Operational Missions. Operating Robotic Exploration missions requires a significant effort, national coordination, and specialised workforce. If a country can operate multiple missions at the same time, its efforts, resources, and specialised workforce are of a different level compared to those of a country that conducts just one mission at a time. This indicator tries to address these efforts, by considering those missions operational between 2011 and 2020. This also includes missions that have been launched earlier than 2011, but that were operational any time until 2020. Scores are assigned based on the maximum number of active missions overlapping during the 2011–2020 decade. Scores for this entry have been attributed as follows:

4 = more than five operational missions overlapping.
3 = multiple operational missions overlapping.
2 = at least one operational mission in 2011–2020, but not overlapping with others.
1 = no mission launched.

Indicator 19: Diversity of Destination. While general interest might be back towards the lunar surface, robotic exploration advocates and mission investigators are targeting a diverse array of destinations across the solar system. Being able to send and operate missions across diverse locations require clear infrastructure, specialised hardware, and workforce. This diversity can be measured as a proxy for scientific prowess, with a special attention to countries planning/launching missions to exotic locations as Jovian moons, comets, asteroids, etc., given the technical difficulties to reach them and the scientific added value of exploring relatively less targeted locations. Scores for this entry have been attributed as follows:

4 = Near Earth Objects, other planets, moons.
3 = Mars.
2 = Moon.
1 = no mission.

Indicator 20: Mission Architecture Variety. Potential mission architectures vary in costs and complexity, based on whether a mission is composed by an orbiter, a lander, or a rover, for instance. Missions could require the deployment of specific data relay systems (as *Chang'e*-4) or have sample-return capabilities. The more different architectures a country can build, the more it showcases its scientific, technological, and engineering prowess. This indicator addresses it by considering Architectures as: Orbiter, Lander, Rover, Sample Return, Data Relay Infrastructure, Drone. Scores for this entry have been attributed as follows:

4 = more than three different architectures employed.
3 = three different mission architectures.
2 = two or less.
1 = no mission.

3.2.1.4 Space Safety and Security

The number of actors and activities in outer space has increased dramatically over the past decades, contributing to making the space environment more "congested, contested and competitive" (US Department of Defense, 2011). The entry of new institutional and private actors, the multiplication of spacecraft launched in orbit, the parallel and largely uncontrolled proliferation of orbital debris, the flourishing of new types of activities as well as the pursuit of space superiority objectives by the military actors of the major powerhouses have all increased the need to properly ensure the safety and security of space activities. Indeed, given the importance that space assets have in modern societies, an increasingly important part of space activities is devoted to developing capabilities and solutions aimed at ensuring their safety and upholding their security. Safety is here intended as the "result of measures precluding inherent malfunction and mitigating the risks of accidental damage that would be caused by or undergone by a space object, including its component parts" (Zarkan Cesari, 2021), while security is defined as "the protection of a space object, including its component parts, against the risk of intentional actions undertaken by external or unauthorized actors" (Zarkan Cesari, 2021).

Two main types of efforts can be broadly identified: Space Situational Awareness (SSA) and counterspace. Consequently, the space safety and security macro-area of our measurement model will be segmented along these two areas. A more detailed description of these two areas is provided hereby.

Space Situational Awareness

Space Situational Awareness (SSA) can be defined as the ability to obtain timely, accurate, comprehensive, and predictive knowledge of our space operating environment with the aim of monitoring hazards from space, determining their risk and, when relevant, mitigating threats posed to both space-based and ground infrastructure (Gasparini & Miranda, 2010; Oltrogge & Alfano, 2019; Weeden, 2017). This ability to characterise the space environment can be acquired through different means, including ground- or space-based optical telescopes, radar systems, lasers, radio-frequency sensors and infra-red sensors. SSA serves both civil and security purposes. In the United States, the national security version of SSA is known as Space Domain Awareness (SDA) (Weeden & Samson, 2022). SDA can be understood as the ability to effectively identify, characterise, and understand any factor, passive or active, associated with the space domain that could affect the security of space operations. Hence, whereas SSA focuses mainly on supporting safety and sustainability efforts, SDA places a key emphasis on detecting and characterizing threats and supporting security efforts, including defensive and offensive operations. Specifically, SDA can serve as critical enabler of offensive counterspace operations (see next section).

The SSA area can be segmented into three main types of activities: Space Surveillance Tracking (SST) of man-made objects, Near Earth Objects (NEO) monitoring, and Space Weather (SWE) monitoring and forecast.

The SST segment is primarily concerned with the detection, production of catalogues and orbit prediction of the artificial objects and debris orbiting the Earth. An SST system is "a network of ground-based and space-based sensors capable of surveying and tracking space objects, together with processing capabilities aiming to provide data, information and services on space objects that orbit around the Earth" (EUSatCen, 2023).

The NEO segment, on the other hand, is concerned with detecting and tracking, mainly through the use of telescopes, natural objects (e.g., asteroids and comets), that could impact Earth. The purpose is "to become aware of the current and future position of near-Earth objects relative to our planet; estimate the likelihood of Earth impacts; assess the consequences of any possible impact; inform relevant parties, e.g., national emergency response agencies; develop methods to deflect any risky asteroids" (ESA, 2023a, 2023b, 2023c, 2023d).

As for Space Weather, the term designates "the physical and phenomenological state of natural space environments. The associated discipline aims, through observation, monitoring, analysis and modelling, at understanding and predicting the state of the Sun, the interplanetary and planetary environments, and the solar and non-solar driven perturbations that affect them, and also at forecasting and nowcasting the potential impacts on biological and technological systems (ESA, 2009).

Consistent with this segmentation, within our measurement model, the "astronomy and science" area has been assessed through the following three indicators:

Indicator 21: Space Surveillance and Tracking. Space Surveillance and Tracking is a critical component of SSA. The deployment of ground-based or space-based systems to track artificial satellites and space debris is both vital to safely operate in space as well as for supporting military missions and objectives in space. SST networks can be based on ground radar or optical observatories, as well as space-based ones; they can also be national in nature or have global coverage through partnerships or tracking ships. This indicator is scored on the basis of following four elements: (a) ground-based optical and laser sensors, (b) ground-based radar sensors, (c) space-based sensors, and (d) global reach. Scores for this entry have been more specifically attributed as follows:

4 = all the elements are present.
3 = three elements present.
2 = two element present.
1 = one or no element is present.

Indicator 22: Near Earth Objects Monitoring. Similar to man-made spacecraft and debris, also natural objects in space can pose a hazard to space and ground infrastructure. NEOs, moreover, can pose a larger, more serious danger to our life on Earth. Being able to identify and track asteroids is a first necessary step to implement planetary defence measures. To assess an actor's capabilities in monitoring NEOs, this indicator considers whether one country has: (a) terrestrial observatories, (b) space-based observatories, (c) launched missions to explore NEOs, (d) institutional Planetary defence-focused agencies or programmes. Scores for this entry have been attributed as follows:

4 = all the elements are present.
3 = three elements are present.
2 = two elements are present.
1 = one or no element is present.

Indicator 23: Space Weather. The study and monitoring of space weather is essential to understand how solar activities interact with our atmosphere and magnetosphere and interfere with the electronics humans have disseminated both on Earth and in orbit. The capability to predict space weather events can spell the difference between a temporary blackout and the complete destruction of space (and terrestrial) hardware. Both terrestrial and orbital observatories have been deployed to monitor solar activities as well as to study earth ionosphere, magnetosphere, etc. To assess an actor's capabilities in monitoring space weather events, this indicator considers whether the actor has: (a) terrestrial solar observatories, (b) dedicated space-based solar observatories, (c) weather satellites with dedicated payloads, (d) institutional space weather-focused agencies/programmes. Scores for this entry have been attributed as follows:

4 = all the elements are present.
3 = three elements are present.
2 = two elements are present.
1 = one or no element is present.

Counter-Space

Over the past two decades, more and more countries have started to emphasise the significance of space applications to support national security objectives. The ensuing use (and increased reliance) on space technologies for conventional warfare and other military activities on Earth has been accompanied by a growing interest in the development of counterspace capabilities. Counter-space, also known as space control, is generally defined as "the set of capabilities or techniques that are used to gain space superiority. Space superiority is the ability to use space for one's own purposes while denying it to an adversary" (Weeden & Samson, 2022). These capabilities can be either offensive or defensive in nature. While the latter are intended to protect space assets from attack, the former aim to prevent the adversary from using their space assets. Offensive capabilities can be more specifically "used to deceive, disrupt, deny, degrade, or destroy any of the three elements of a space system: the satellite, the ground system, or the communication links between them" (Weeden & Samson, 2022).

Counter-space capabilities are based on a variety of technologies, both ground and space-based, which can be divided in four main categories: (a) Direct-Ascent, (b) Direct Energy, (c) Co-Orbital, and (d) Electronic Warfare. As reported by Secure World Foundation:

- Direct Ascent designates "weapons that use ground-, air-, or sea-launched missiles with interceptors that are used to kinetically destroy satellites through force of impact, but are not placed into orbit themselves.
- Co-orbital designates "weapons that are placed into orbit and then manoeuvre to approach the target to attack it by various means, including destructive and non-destructive."
- Directed Energy designates "weapons that use focused energy, such as laser, particle, or microwave beams to interfere or destroy space systems."
- Electronic Warfare indicates "weapons that use radiofrequency energy to interfere with or jam the communications to or from satellites" (Weeden & Samson, 2022).

Within our measurement model, the "counter-space" area has been assessed through the following indicators:

Indicator 24: R&D of Counter-Space Capabilities. This indicator aims to assess whether the considered actors have undertaken R&D activities across four main areas of counter-space: Direct-Ascent, Direct Energy, Co-Orbital, and Electronic Warfare. Today no country has disclosed R&D activities across the four areas. This indicator has been scored on the basis of Secure World Foundation's Global Counter Space Capabilities Report.[1] Scores for this entry have been attributed as follows:

4 = research in four areas or more.
3 = research in three areas.
2 = research in two areas.
1 = research in one area or less.

Indicator 25: Testing of Counter-Space Capabilities. This indicator seeks to evaluate whether the considered actors have undertaken the testing of counter-space capabilities across the four main areas (Direct-Ascent, Direct Energy, Co-Orbital, and Electronic Warfare). Today countries do not publicise the testing of counter-space capabilities, although tests (if not obvious, as in the case of Direct-Ascent tests) can be deducted by, for instance, satellites' behaviour. The scores of this entry

[1] The report "Global Counter Space Capabilities. An Open Source Assessment", is an annual, internationally acknowledged, publication of Secure World Foundation, a private operating foundation that promotes cooperative solutions for space sustainability and the peaceful uses of outer space. The report compiles and assesses publicly available information on the counterspace capabilities being developed by multiple countries across five categories: direct-ascent, co-orbital, electronic warfare, directed energy, and cyber. It assesses the R&D, testing and operational use of these capabilities for each country, along with their potential military utility (Weeden & Samson, 2022).

have been produced on the basis of Secure World Foundation's Global Counter Space Capabilities Report. Scores for this entry have been attributed as follows:

4 = testing in four areas or more.
3 = testing in three areas.
2 = testing in two areas.
1 = testing in one area or less.

Indicator 26: Deployment of Counterspace Capabilities. This indicator assesses whether the considered actors have undertaken the deployment of counter-space capabilities across the four main areas (Direct-Ascent, Direct Energy, Co-Orbital, and Electronic Warfare). Today countries do not publicise the deployment of counter-space capabilities. Therefore, this indicator has been scored on the basis of Secure World Foundation's Global Counter Space Capabilities Report. Scores for this entry have been attributed as follows:

4 = capabilities in four areas or more deployed.
3 = capabilities in three areas deployed.
2 = capabilities in two areas deployed.
1 = capabilities in one area or less deployed.

3.2.1.5 Enabling and Support

Enabling and support is a transversal macro-area covering all the different types of activities, technical capabilities and infrastructure required to enable and sustain the conduct of any space mission. The most evident enabler is space transportation, which is the gateway for any orbital activity and the deployment of any asset in space. From Sputnik to Hubble, passing through Apollo and the modules of the International Space Station (ISS), space programmes worldwide have inescapably relied on the availability of launch vehicles to reach and exploit outer space (Aliberti & Tugnoli, 2015).

Beyond the transportation system itself, also critical for the successful conduct of space activities are all the infrastructural elements that support the launch, deployment and operations of space assets. These include the launch facilities, the mission command and control centres, the different land-, sea- and space-based tracking stations and, more broadly, all facilities performing telemetry, tracking, and control (TT&C) functions to ensure the successful operation of a spacecraft.

Specific types of technical capabilities support activities such as rendezvous, docking, berthing and inspection operations, robotic manipulation of spacecraft, in-orbit assembly, in-orbit refuelling—and technologies can also be included in this macro-area. Within our measurement model, the enabling and support macro-area has been divided in three main areas: Space transportation, ground operations, space operations and tech demonstrators. A description of these three areas is provided hereby.

Space Transportation

Space transportation provides the foundational ability to deliver payloads into space. SLVs place satellites in orbit to deploy, sustain, augment, or reconstitute assets in orbit in support of military, civilian, or commercial purposes (JCS, 2020). For any nation with a minimum of strategic ambition in space, space transportation systems (generally orbital carrier rockets) represent a strategic good—or more precisely a strategic enabler: not a goal in itself, but a *conditio sine qua non* for the conduct of their military, civil or commercial space efforts, and thus the foundation for any comprehensive space policy and for the full utilisation of space assets. Understandably, the establishment of indigenous launch systems has been generally regarded as a crucial priority that initially responded to security needs as well as to considerations of strategic independence and national pride, but which was ultimately instrumental in providing a wide range of socio-economic benefits through the exploitation of space assets (Aliberti & Tugnoli, 2015).

However, even if the equation *no launcher, no space programme* provides a rather straightforward rationale for the establishment of national launch capacities, attaining such a goal is no easy undertaking. Quite to the contrary, the development and maintenance of an indigenous space launch system and related infrastructure is a capital-intensive process requiring extremely advanced technical capabilities and large financial investment. In view of that, it is not surprising that only a dozen countries have been able to establish a basic launch capacity to reach lower altitude orbits. Multi-destination capabilities to reach the geostationary orbit (GTO), low and medium earth orbits and deep space are even more rare (Aliberti & Tugnoli, 2015).

Within our measurement model, the "space transportation" area has been assessed through the following indicators:

Indicator 27: Launch Cadence. The number of launches conducted nationally can be a proxy for the dynamism of a nation and its actions in space, as well as a way to measure the level of international commercial satellite launch market captured by the same nation. The indicator is scored by considering if a space actor has launched: (a) less than once per year, (b) less than once every two months, (c) less than once every month, (d) or more than once per month on average in the period 2011–2020. Scores for this entry have been attributed as follows:

4 = More than 120 launches in the period 2011–2020.
3 = Between 61 and120 launches in the period 2011–2020.
2 = Between 11 and 60 launches in the period 2011–2020.
1 = Between 0 and 10 launches in the period 2011–2020.

Indicator 28: Launchers' Variety. A variety of launchers and launch service providers grants resilience and flexibility to national actors, and different payloads sizes, timelines and reachable orbits enable more space activities. This indicator divides the launchers in four classes: (a) small-lift, (b) medium-lift, (c) heavy-lift, and (d) superheavy-lift launch vehicles. Scores are assigned based on active (i.e., that were effectively used) launchers in 2011–2020. Scores for this entry have been attributed as follows:

4 = three or more launch vehicle classes covered.
3 = two classes covered.
2 = one class covered.
1 = none.

Indicator 29: Launchers' Performance. While satellites are becoming lighter thanks to technology miniaturisation, heavy-lift capabilities remain still important to secure launches to GEO, to build large infrastructures in space, conduct Deep Space missions as well as crewed activities. A comparison of countries based on publicly available information on Kg-to-LEO and Kg-to-GTO capability of their active launchers can be an indicator of the type of activities a country could be enabled to conduct in space. This indicator is measured by comparing single launchers' kg-to-LEO and Kg-to-GTO to the average performance of the most powerful active launchers—with active launchers being those that have been used in the period 2011–2020 (Falcon Heavy, Long March-5, Proton-M, Ariane-5, H-IIB, GSLV Mk.3, Electron, Shavit-2, Naro-1). The average Kg-to-LEO considered is 18000 kg, while the average Kg-to-GTO considered is 11830 kg. Scores for this entry have been attributed as follows:

4 = double or more the average Kg-to-LEO/Kg-to-GTO.
3 = above the average Kg-to-LEO/Kg-to-GTO.
2 = below the average Kg-to-LEO/Kg-to-GTO.
1 = micro launcher/no launcher.

Indicator 30: Launchers' Reliability. Autonomous national space programmes and a thriving launch market require reliable launchers. Countries have been scored based on the success rate of their launches in 2011–2020. The success rate of countries that have launched less than once per year (Israel, South Korea) has not been considered, and such countries have been scored the lowest, given the low launch cadence that cannot fully ensure reliability. Scores for this entry have been attributed as follows:

4 = Success rate equal or above to 95%
3 = Success rate between 90 and 95%
2 = Success rate between 80 and 90%
1 = Success rate below 80%, limited number of launches, or no launcher.

Ground Operations

Essential for the proper deployment and operations of any spacecraft are the launch site facilities from which payloads are delivered into space as well as the Telemetry, Tracking and Command TT&C infrastructure, which provides a connection between the satellite itself and the facilities on the ground. The TT&C infrastructure hence ensures the ability of the spacecraft to successfully achieve its purpose (Guest, 2013). As the name suggests, the TT&C infrastructure "performs three specific tasks:

- Telemetry. The collection of information on the health and status of the entire satellite and its subsystems and the transmission of this data to the command segment on the ground. This requires not only a telemetry system on the spacecraft but also for a global network of ground stations around the world to collect the data, unless, of course, the application satellite network includes inter-satellite links that are capable of relaying the data to a central collection point.
- Tracking. The act of locating and following the satellites to allow the command segment to know where the satellite is and where it is going. Again, this requires a ranging system on the spacecraft and a data collection network on the ground that allows this ranging and tracking function to work.
- Control. The reception and processing of commands to allow the continuing operation of the satellite in order to provide the service of interest. Once again, an appropriate ground system is required (Guest, 2013).

Within our measurement model, the "ground operations" area has been assessed through the following indicators:

Indicator 31: TT&C Infrastructure. Satellite transmissions and TT&C capabilities are a fundamental part of space activities. While all countries with a spacecraft in orbit may have some expertise in communicating with their assets, the capability to communicate with scientific, human spaceflight, deep space/interplanetary missions require more advanced communication systems. These types of missions require advanced technology not available on the market, building ground stations outside of the state's territory, advanced data relay systems, huge coordination across various missions, and a specific expertise in managing radiofrequency spectrum. Scoring the various countries based on their deep space network can be a proxy to compare the countries' prowess in TT&C technologies. This indicator hence considers whether the considered actor possesses: (a) a Deep Space Network, (b) a Space-based Data Relay System, (c) Tracking Vessels that reduce terrestrial constraints. Scores for this entry have been attributed as follows:

4 = all the three elements possessed.
3 = two of the elements possessed.
2 = just one of the elements possessed.
1 = no element possessed (basic or no TT&C capabilities).

Indicator 32: Launch Facilities. Having multiple national launch facilities enables more space activities, as owning a limited number of launch pads can be a bottleneck for launches. A country that has multiple launchpads require relevant infrastructures and talent to manage them. This indicator addresses this by comparing countries based on the number of active spaceports from where orbital-class rockets have been launched during the period 2011–2020. Scores for this entry have been attributed as follows:

4 = three or more active orbital spaceports in 2011–2020.
3 = two active orbital spaceports in 2011–2020.
2 = one active orbital spaceport in 2011–2020.
1 = no active spaceport in 2011–2020.

3.2.1.6 Space Operations

Space operations encompasses all the activities undertaken to launch, manoeuvre, configure, operate, and sustain on-orbit spacecraft (Joint Chiefs of Staff, 2020). They include launch operations, spacecraft operations such as TT&C, manoeuvring, monitoring state-of-health, and maintenance sub-functions as well as payload operations to monitor and control that the satellite payload collects data or provide the intended capability (Joint Chiefs of Staff, 2020). Space operations, also include Rendezvous and Proximity Operations (RPO), which are specific processes where two resident space objects are intentionally brought close together for the inspection, repair, replacement, upgrade or removal of the spacecraft. RPO technologies and activities are currently developed by governmental, military, and private actors alike for applications like Active-debris removal (ADR) or life extension services.

Within our measurement model, the "space operations" area has been assessed through the following indicators:

Indicator 33: Active-Debris Removal (ADR) Capabilities. Rendezvous and Proximity Operations (RPO) technologies are currently pursued by governmental, military, and private actors. This set of technologies and standards include capabilities as rendezvous, docking, berthing, inspection, robotic manipulation of spacecrafts. Active-Debris Removal (ADR) is a specific type of space operations linked to RPO capabilities and technologies. While no operational service has already been deployed, technological readiness level is getting closer to that. This indicator is therefore measured by high-level scoring TRL, namely: available technology, in-orbit demonstration, service deployed. RPO technologies have significance for ADR applications, so they will be counted as available technology. This indicator addresses whether countries have developed technology related to ADR, validated it in-orbit, or actually deployed mature ADR service. Scores for this entry have been attributed as follows:

4 = service deployed.
3 = in-orbit demonstration.
2 = technology available.
1 = no tech.

Indicator 34: Life Extension Services. Rendezvous and Proximity Operations (RPO) include capabilities such as rendezvous, docking, berthing, inspection, robotic manipulation of spacecrafts. Life extension services are application of such technologies and include operations as station keeping, orbital manoeuvring, in-orbit assembly, inspection, repair, replacement, upgrade and refuelling of spacecraft (e.g., fuels, fluids, cryogens). Some of these capabilities have already been tested in connection with human spaceflight (e.g., progress capsules boosting or refuelling the ISS), while endeavours are undergoing to develop commercially viable solutions with private initiatives slowing reaching maturity. This indicator addresses whether countries have developed technology related to Life Extension services, validated it in-orbit, or actually deployed mature Life Extension services. RPO technology has been considered as potentially transferable to Life Extension services. Scores for this entry have been attributed as follows:

4 = all the technologies described are present.
3 = two of the technologies described are present.
2 = one of the technologies described is present.
1 = none of the technologies described is present.

Indicator 35: In-space Deployment Capabilities. The capability to deploy spacecrafts and hardware from orbing platforms is a relevant technology that can facilitate access to space. This indicator considers (1) Deployers from the ISS, (2) space transfer/transport vehicles as tugs or spaceplane, and (3) the capability to integrate multiple and different spacecraft both in rideshare missions or as piggyback to a primary spacecraft, when a significant number of spacecrafts is involved (e.g.: Europe's SSMS, SpaceX's Transporter missions). Scores for this entry have been attributed as follows:

4 = all the technologies/capabilities described are present.
3 = two of the technologies/capabilities described are present.
2 = one of the technologies/capabilities described is present.
1 = none of the technologies/capabilities described is present.

Tech Demonstrators

Since the beginning of the space age, a large number of experimental spacecraft that do not fall under specific categories considered so far has been deployed in space. As NASA puts it, these missions are nonetheless fundamental as they "bridge the gap between need and means, between scientific and engineering challenges and the technological innovations needed to overcome them, between early proof-of-concept tests and the final infusion of cost-effective, revolutionary new technologies into […] government and commercial space missions" (NASA, 2021). Technology precursors and demonstrators have been launched by space agencies, military actors, commercial companies and even by universities and amateurs.

Within our measurement model, the "tech demonstrators" area has been assessed through the following indicators:

Indicator 36: Number of Tech Demonstrator Missions. A country whose actors can afford to launch experimental spacecrafts is well positioned to have or gain an edge over the other space actors in the medium-to-long-term. Accordingly, this indicator addresses the number of technology demonstrators launched in 2011–2020 and operated by each country. Scores for this entry have been attributed as follows:

4 = more than 100.
3 = between 21 and 99.
2 = between 2 and 20.
1 = one or none.

Indicator 37: Nature of Tech Demonstrators. The scope and nature of a space tech demonstrator in orbit can be a valid indicator displaying the level of dynamism in a country's space sector. Promising countries will have not only the military or the government involved in technology demonstration, but also commercial actors and academic institutions. This indicator addresses this dynamism, by scoring the various countries based on the nature of the actors involved in the demonstrations: Amateur, Commercial, Educational, Governmental, Military. Scores for this entry have been attributed as follows:

4 = four types of actors or more.
3 = three types of actors.
2 = two different types of actors.
1 = just one type of actor or no demonstrator.

3.2.2 *Soft Capacity*

With respect to the soft capacity sub-dimension, it is possible to identify a number of key features which are considered to assess whether space is effectively leveraged or embedded in the national infrastructure or the implementation of public policies by national institutions. From an overall perspective, space activities can be said to support the fulfilment of economic, societal, political and strategic goals within two major policy domains: "socio-economic policies" and "foreign and security policies." For both macro-areas, Table 3.2 presents a specific set of indicators. Importantly, macro-areas have been broken down into four and three areas, respectively.

3.2.2.1 Socio-Economic Policies

The possibilities offered by space through its technologies and applications are enormous. The integration of Earth observation, communications, and navigation services

Table 3.2 Measuring soft capacity: macro-areas, areas, and entries

Macro-area	Area	Entry
Socio-economic policies	Environment and resources	38: Agriculture
		39: Other natural resources
		40: Meteorology
		41: Environment
		42: Climate change
	Infrastructures	43: Infrastructures management
		44: Energy
		45: Mobility
	Development and growth	46: Rural and urban development
		47: Scientific and technological innovation
		48: Industrial development
		49: Market development and commercialisation
	Civil society	50: Health
		51: Education
		52: Entertainment and citizen services
		53: National identity
Foreign and security policies	Security	54: National security policies
		55: Surveillance and monitoring
		56: Disaster prevention and management
	Defence	57: National military strategy
		58: Prevention and deterrence
		59: Military operations
	Foreign policy	60: Space for diplomacy
		61: Space diplomacy
		62: Foreign aid and international initiatives
		63: International prestige

into our daily lives boosts economic growth and generates employment. Even more importantly, it also accrues a myriad of tangible benefits for individual citizens and the society as a whole.

The socio-economic benefits associated with space activities are today well known and documented in a number of studies regularly undertaken by research institutes, universities, consulting companies, and international organisations like the OECD, among others. In general, these studies underline that the role of space systems in national economy has substantially grown and that the space infrastructure today provides a broad range of services that are essential for a variety of strategic domains and economic sectors, from security and defence to agriculture, energy and banking, just to name a few (Moranta et al., 2018).

Beyond delivering turnkey services to end-users in a plethora of fields, the use of space-based solutions contributes to the fulfilment of a variety of national policy objectives. Space assets have become instrumental for the implementation of many public sectorial policies, both directly (i.e., when space-based solutions are used directly by public institutions to achieve national objectives) and indirectly (i.e., when the use of space-based solutions by actors/users of target sectors supports the achievement of national objectives) (Moranta et al., 2018). Examples of these contributions include, among others, the support offered by communications, remote sensing and PNT data to the implementation of national policies intended to, for instance, foster agricultural productivity; support the sustainable exploitation of fisheries resource; enable competitive, sustainable, and safe transport services; bridge the digital divide between urban areas and remote locations; or monitor environmental resources and climate change. Additionally, the uptake of space technology plays a huge part in innovation policies, laying the foundations for the cross-pollination of space technologies with ground technologies, and the development of new services in which space systems are key enablers (e.g., 5G networks, air traffic management, smart energy grids, and autonomous vehicles). More broadly, the space sector often serves as a "driver for growth and innovation," generating considerable benefits at the macroeconomic level (OECD, 2012).

The above are only a few of the many contributions offered by space activities to the fulfilment of national objectives. In fact, space capabilities are used or could be used, in a vast majority of policy areas that require public intervention, ranging from urban planning and cultural heritage characterisation to telemedicine and tele-education applications. The magnitude of benefits enabled by space assets varies among policy areas but can be critical when space assets provide a particularly efficient, and sometimes irreplaceable, means to achieving policy objectives (Moranta et al., 2018).

The "socio-economic policies" macro-area has been segmented in four main areas, namely: environment and resources, infrastructure, development and growth and civil society. A description of these three areas is provided hereby.

Environment and Resources

Space technologies and services provide significant contributions to managing the environment and its natural resources. Remote sensing satellites, for instance, provide critical data that can be used for natural resources inventory, surveying and mapping, hydrology, geology, mineralogy, ecology, and the monitoring of land cover/use. Space applications also offer unvaluable tools to the agricultural sector. For instance, by providing accurate data and analysis of the soil, drought, rainfall and crop development, space assets can support farmers, agronomists, food manufacturers and policymakers in increasing productivity and profitability of the agricultural sector (UNOOSA, 2023c).

Similarly, Earth observation and meteorological satellites are vital for weather forecasting, providing a unique perspective and sets of data that support predictions, weather modelling, early warnings, etc. They help meteorologists in identifying potentially dangerous weather situations and in issuing timely forecasts and warnings to emergency services and local authorities (EUMETSAT, 2023). Meteorological satellites also provide critical inputs to monitoring and forecasts of air quality and collecting continuous, long-term global observations of the oceans.

More broadly, space-based data provide essential information on the past, current, and future state of climate change. Remarkably, half of the Essential Climate Variables (ECVs), which are necessary to monitor climate change, directly come from satellite data. In addition, satellite data provide the scientific evidence required for climate-related policies and can help decisionmakers to set goals to reduce the environmental footprint and establish adaptation, mitigation, and resilience measures. They widely contribute to scientific research conducted by international institutions such as the Intergovernmental Panel on Climate Change (IPCC).

Within our measurement model, the "environment and resources" area has been assessed through the following indicators:

Indicator 38: Agriculture. This indicator aims to assess the integration of space technologies within the agricultural sector of the considered actor the use of Earth observation and PNT data and services for precision agriculture, soil monitoring, crop production forecasting, crop intensification, rotation and cropping system analysis, etc.

Scores for this indicator have been attributed by country experts invited to answer the survey question: "on a scale from 1 to 4, how often would you say that space assets are used to support the agricultural sector in the country? Answers range from 1 (never) to 4 (very often). A minimum of 3 responses have been submitted for each country. Answers from multiple respondents have been averaged to form a single score.

Indicator 39: Other Natural Resources. This indicator measures the extent to which space technologies and applications are used to support conservation and sustainable management of natural resources by the considered actor. Examples include the use of remote sensing or PNT data for forestry inventory and vegetation monitoring, land use/land cover mapping, wetland inventory and conservation plans, bio- resource characterisation, water reservoir capacity evaluation, ocean productivity assessments and identification of optimal fishing locations, just to name a few.

Scores for this indicator have been attributed by country experts invited to answer the survey question: "on a scale from 1 to 4, how often would you say that space assets are used to support natural resources management (e.g., forestry, fishing, mining, etc.) in the country?" Answers range from 1 (never) to 4 (very often). A minimum of 3 responses have been submitted for each country. Answers from multiple respondents have been averaged to form a single score.

Indicator 40: Meteorology. This indicator aims to assess the level of integration of space assets and applications in the weather forecasting and meteorological activities of the considered actor by e.g., providing data on air quality and atmospheric composition, ozone layer and ultra-violet radiation, solar radiation, etc.). Scores for this indicator have been attributed by country experts invited to answer the survey question: "on a scale from 1 to 4, how often would you say that space assets are used to support meteorology and weather forecasting activities in the country?" Answers range from 1 (never) to 4 (very often). A minimum of 3 responses have been submitted for each country. Answers from multiple respondents have been averaged to form a single score.

Indicator 41: Environment. This indicator aims to measure how extensively space assets are used to support environmental monitoring and inform environmental action (e.g., through the provision of data on wildlife and vegetation for biodiversity and ecosystems protection) by the considered actor. Scores for this indicator have been attributed by country experts invited to answer the survey question: "on a scale from 1 to 4, how often would you say that space assets are used to support environmental monitoring and/or protection (biodiversity and ecosystems) in the country?" Answers range from 1 (never) to 4 (very often). A minimum of 3 responses have been submitted for each country. Answers from multiple respondents have been averaged to form a single score.

Indicator 42: Climate Change. This indicator aims to assess the extent to which space assets and space-based data are used to support climate change monitoring through the delivery of climate variables, contribute to scientific research conducted by recognised international institutions such as the IPCC, and inform mitigation and adaptation policies in the country by providing consistent and authoritative evidence about climate change. Scores for this indicator have been attributed by country experts invited to answer the survey question: "on a scale from 1 to 4, how often would you say that space assets are used to support climate change monitoring and/or mitigation/adaptation policies in the country?" Answers range from 1 (never) to 4 (very often). A minimum of 3 responses have been submitted for each country. Answers from multiple respondents have been averaged to form a single score.

Infrastructures

Whether it is infrastructure management, urban planning, mobility, or energy, all these fields can improve substantially through the use of space solutions. Major communications networks, banking systems, financial markets, and power grids all depend heavily on GNSS for precise time synchronisation. With space-based communication, satellite imagery, precise timing and tracking services, space can provide solid means to manage, monitor and make infrastructures, more secure, resilient, and sustainable.

In the energy sector, space assets enable tasks such as preventive or immediate maintenance, risk or status assessment, surveying of large areas, spot pipe leakage, and many more activities down to counting vehicles or measuring the filled volume of tankers/tanks.

Space assets also provide vital tools for the mobility sector, as they support precise monitoring and tracking of assets, as well as traffic management of vehicles across different mediums (land, sea, or air). Space can be used to enhance navigation at sea, in air, or in remote areas or contribute to the creation of "smart cities dashboard solutions" for enhancing the quality of urban planning and situational awareness through the integration of Earth observation, navigation and telecommunication services.

Within our measurement model, the "infrastructure" area has been assessed through the following indicators:

Indicator 43: Infrastructures Management. This indicator aims to assess the extent to which space assets and space-based data contribute to supporting infrastructure monitoring and management in the area of e.g., construction, logistics, banking systems and financial markets by the considered actor.

Scores for this indicator have been attributed by country experts invited to answer the survey question: "on a scale from 1 to 4, how often would you say that space assets are used to support infrastructure management (e.g., construction, logistics, finance, etc.) in the country?" Answers range from 1 (never) to 4 (very often). A minimum of 3 responses have been submitted for each country. Answers from multiple respondents have been averaged to form a single score.

Indicator 44: Energy. This indicator aims to assess the extent to which space assets and space-based data are used to support the energy sector by the considered actor. Examples include the utilisation of space-based solutions for site identification, pipelines and grids timing and synchronisation, network management, etc.

Scores for this indicator have been attributed by country experts invited to answer the survey question: "on a scale from 1 to 4, how often would you say that space assets are used to support the energy sector (e.g., site identification, pipelines and grids timing and synchronisation, etc.) in the country?" Answers range from 1 (never) to 4 (very often). A minimum of 3 responses have been submitted for each country. Answers from multiple respondents have been averaged to form a single score.

Indicator 45: Mobility. This indicator aims to assess the extent to which space assets and space-based data are used to support the mobility sector by the considered actor. Examples include the utilisation of space-based solutions for monitoring and tracking assets, managing traffic across different mediums, enhancing safety in congested area or increasing efficiencies and reduced costs in surveying roads.

Scores for this indicator have been attributed by country experts invited to answer the survey question: "on a scale from 1 to 4, how often would you say that space assets are used to support transport and/or mobility (e.g., land, water, air navigation, traffic monitoring, goods tracking, etc.) in the country? Answers range from 1 (never) to 4 (very often). A minimum of 3 responses have been submitted for each country. Answers from multiple respondents have been averaged to form a single score.

Development and Growth

The 2022 projections by the United Nations Department of Economic and Social Affairs (UNDESA, 2022) indicate that the global population will grow to around 8.5 billion in 2030, 9.7 billion in 2050 and 10.4 billion in 2100. The accompanying urbanisation prospects posing clear challenges with housing, water, sanitation, electricity, pollution and transportation. As reported by UNOOSA, "space-based technologies provide unique tools for planning socially and environmentally sustainable human settlements. Central government policymakers, mayors, city planners, engineers and landscape architects are among those who use remote sensing tools that measure and monitor existing patterns of land use and infrastructure development. Not only does this data inform decision makers about current urban projects, complex models can also be constructed to predict future trends in human settlements and urbanization" (UNOOSA, 2023d).

Besides supporting sustainable urban and rural development, space can more broadly provide substantial contribution to sustainable development by closing the traditional infrastructure gaps in developing countries and contributing to the emergence of digitally empowered societies. Space is also an engine of economic growth. Space programme can indeed be aptly leveraged as an enabler of economic expansion in terms of macro- economic production functions, particularly in terms of scientific and technological innovation as well as industrial upgrading, market creation, and development of a skilled workforce. The positive spillover effect of space technologies has been validated in a variety of studies and through a variety of approaches. The OECD, for instance, has categorised the diversity of economic impacts derived from space activities in different segments, namely, new commercial products and services (including "indirect industrial effects" from space industry contracts, meaning new exports or new activities outside the space sector), productivity/efficiency gains in diverse economic sectors, national and regional economic growth and cost avoidances (e.g., through disaster prevention) (OECD, 2011).

Within our measurement model, the "development and growth" area has been assessed through the following four indicators:

Indicator 46: Rural and Urban Development. This indicator aims to assess the extent to which space assets and space-based data contribute to supporting urban and/or rural development by the considered actor through the provision of relevant data, services and know-how informing decision-making or the implantation of policy actions.

Scores for this indicator have been attributed by country experts invited to answer the survey question: "on a scale from 1 to 4, how often would you say that space assets are used to support urban and/or rural development (e.g., survey and mapping, development plans, wasteland management, etc.) in the country?" Answers range from 1 (never) to 4 (very often). A minimum of 3 responses have been submitted for each country. Answers from multiple respondents have been averaged to form a single score.

Indicator 47: Scientific and Technological Innovation. This indicator aims to assess the extent to which the research, technologies, processes, data, and know-how generated by the conduct space activities contribute to scientific and technological innovation in other economic sectors.

Scores for this indicator have been attributed by country experts invited to answer the survey question: "on a scale from 1 to 4, how often would you say that space activities contribute to scientific and/or technological innovation in other national economic sectors?" Answers range from 1 (never) to 4 (very often). A minimum of 3 responses have been submitted for each country. Answers from multiple respondents have been averaged to form a single score.

Indicator 48: Industrial Development. This indicator aims to assess the extent to which the research, technologies, processes, data, and know-how generated by the conduct space activities contribute to the development of a strong and competitive industrial base in the country in both the aerospace sector and beyond.

Scores for this indicator have been attributed by country experts invited to answer the survey question: "on a scale from 1 to 4, how often would you say that space activities contribute to industrial development and the emergence of a robust industrial base in the country?" Answers range from 1 (never) to 4 (very often). A minimum of 3 responses have been submitted for each country. Answers from multiple respondents have been averaged to form a single score.

Indicator 49: Market Development and Commercialisation. This indicator aims to assess the extent to which the research, technologies, processes, data, and know-how generated by the conduct space activities contribute to stimulating (new) markets development and commercial activities, based both on space-based data and on sectoral spin-ins and spin-offs.

Scores for this indicator have been attributed by country experts invited to answer the survey question: "on a scale from 1 to 4, how successful would you say that the country is in using space to stimulate market development and commercial activities?" Answers range from 1 (not successful at all) to 4 (very successful). A minimum of 3 responses have been submitted for each country. Answers from multiple respondents have been averaged to form a single score.

Civil Society

As an important growth driver, space activities not only enable industrial development and economic growth. They also yield a variety of tangible and intangible benefits to large segments of the civil society. Relevant quality-of-life benefits can be observed in a number of areas (e.g. health, education, transportation, consumer goods, information technology and entertainment services).

Essential areas such as health care are becoming more flexible and remotely accessible thanks to satellite systems. In this field, the use of Earth observation, telecommunication, navigation, and scientific research in micro-gravity environment have greatly enhanced medical research, remote diagnosis and treatment, public health logistics, and emergency management. As reported by UNOOSA, "information from remote sensing technologies is, for instance, applied to study the epidemiology of infectious diseases. Data is used to monitor disease patterns, understand environmental triggers for the spread of diseases, predict risk areas and define regions that require disease-control planning. This tele-epidemiology is of particular relevance in developing countries, where infectious diseases remain among the top causes of death" (UNOOSA, 2023d).

Furthermore, space-based technologies like satellite communications systems can help bridging the education access gap in remote and rural communities through the creation of virtual classrooms and the provision of video-conferencing tools. Beyond facilitating access to educational programme, "space also plays an inspirational role in education. Classes on space topics often spark students' curiosity and imagination and encourage youth of both genders to pursue careers in scientific and technical disciplines (UNOOSA, 2023e).

More broadly, space can be used as a powerful source of inspiration for the entire nation and as a political tool to boost national identity, increase national pride and confidence and reinforce social cohesiveness. Because of the complexities and investments associated with the conduct of space activities, citizens can grow excitement about the first astronaut mission of their country, be inspired by space exploration endeavours, and develop pride from their national programme. These opportunities are of great value and should not be underestimated.

Within our measurement model, the "civil society" area has been assessed through the following indicators:

Indicator 50: Health. This indicator aims to assess the extent to which space assets contribute to supporting the health sector and the implementation of health policies by the considered actor. Examples include access to healthcare via space-enabled telemedicine services or monitoring disease patterns through space-based tele-epidemiology services. Scores for this indicator have been attributed by country experts invited to answer the survey question: "on a scale from 1 to 4, how often would you say that space assets are used to support the health sector (e.g., telemedicine and tele-epidemiology services) in the country?" Answers range from 1 (never) to 4 (very often). A minimum of 3 responses have been submitted for each country. Answers from multiple respondents have been averaged to form a single score.

Indicator 51: Education. This indicator aims to assess the extent to which space assets contribute to supporting the education sector and the implementation of health policies by the considered actor. Supporting access to education for rural communities via space-enabled tele-education services is one example. Space programmes can be also targeted as educational activities, providing data or hands-on experience for students. Scores for this indicator have been attributed by country experts invited to answer the survey question: "on a scale from 1 to 4, how often would you say that space assets are used to support the education sector (e.g., remote learning services) in the country?". Answers range from 1 (never) to 4 (very often). A minimum of 3 responses have been submitted for each country. Answers from multiple respondents have been averaged to form a single score.

Indicator 52: Entertainment and Citizen Services. This indicator aims to assess the extent to which space assets are used to enable the provision of entertainment and other services (such as broadcasting, internet, GNSS or Earth observation-enabled applications, etc.) bringing direct and quality-of-life benefits to large segments of the civil society. While citizens might not be necessarily aware while using space-based services, policymakers ought to know what benefits can be derived from space applications in order to ensure appropriate policies, be them for the procurement of additional space assets or for a better distribution of data to stimulate market uptake and the development of the space applications downstream sector.

Scores for this indicator have been attributed by country experts invited to answer the survey question: "on a scale from 1 to 4, how often would you say that space assets are used to provide entertainment and other citizen services (e.g., broadcasting, internet services, GIS, etc.) in the country?" Answers range from 1 (never) to 4 (very often). A minimum of 3 responses have been submitted for each country. Answers from multiple respondents have been averaged to form a single score.

Indicator 53: National Identity. This indicator aims to assess the extent to which space programmes contribute to creating or reinforcing national identity, unity, pride and social cohesiveness within the considered country. Scores for this indicator have been attributed by country experts invited to answer the survey question: "on a scale from 1 to 4, how successful would you say that the country is in using space to create/boost national identity and social cohesion?" Answers range from 1 (not successful at all) to 4 (very successful). A minimum of 3 responses have been submitted for each country. Answers from multiple respondents have been averaged to form a single score.

3.2.2.2 Foreign, Security and Defence Policies

Beyond providing substantial contributions to the definition, implementation and fulfilment of socio-economic policies, space can be a potent tool for meeting the national objectives set forth by foreign, security and defence policies. From a foreign policy and diplomatic perspective, space assets can play a remarkable role in addressing important global issues such as, for instance, the fight against climate change, sustainable development, capacity building and humanitarian aid to populations, countries and regions facing natural or man-made disasters.

Looking at the domain of security, national space assets like PNT, remote sensing and telecommunication satellites have become instrumental for supporting a variety of activities both within and outside the national territory including, among others, civil protection and police mission, maritime security (e.g., traffic monitoring, surveillance of illegal activities, Search & Rescue missions), border surveillance, crisis prevention as well as disaster management. As highlighted by UNOOSA (2023b), "space-based technologies can contribute to all phases of the disaster management cycle, including prevention, preparedness, early warning, response and reconstruction. Before a disaster takes place, remotely sensed data provides information for systems and models which can predict disasters and provide early warnings. Satellites are also reliable and rapid in communication, observation and positioning tools, which become particularly vital to relief and recovery operations when ground-based infrastructure is damaged."

Similarly, the role played by space in the realm of defence and in support of military operations is invaluable. The main functions of interest to the military encompasses intelligence, surveillance and reconnaissance (which includes Earth observation, signal intelligence, early warning and meteorology); satellite communications; positioning, navigation and timing; space surveillance (Bataille & Messina, 2020). For instance, PNT systems nowadays offer the military around the world with irreplaceable data for troop movement, force tracking, and precision munition delivery (DIA, 2022). Similarly, satellite communications are a core competence for military command and control, enabling the quick transmission of orders and critical intelligence, and enhancing the flexibility of modern armies (Bataille & Messina, 2020). It should therefore come as no surprise that the US Department of Defense recognises space "as a priority domain of national military power that underpins multi-domain joint and combined military operations to advance national security," and also highlights the need to "conduct operations in, from, and to space, and deliver advanced space capabilities to deter conflict and, if deterrence fails, to counter and defeat aggression" (DoD, 2022).

Consistent with the above, the "foreign and security policies" macro-area has been segmented in three main areas: security, defence and foreign policy. A description of these three areas is provided hereby.

Security

Security is a top priority of policymakers and ensuring it is one of the main responsibilities of states. Space systems and solutions have long supported security activities, but the interaction between the two dimensions has been growing in recent decades and is now openly asserted. Both traditional and emerging spacefaring nations are launching satellites, or using the data of private space companies, to protect their territory and people.

Security can be conceived along three main pillars, which all benefit from space solutions: law enforcement, emergency management and civil security (Bataille & Hrozensky, 2022). Space systems and services are particularly valuable thanks to their coverage, speed or precision. Overall, the three main types of space applications (remote sensing; satellite communications; PNT) are useful for security actors. For instance, they enable early risk identification and early warning by deepening understanding, detecting major threats, mapping their potential consequences, and preparing for the emergency response in the near-to-medium term. Through the use of Earth observation data, for instance, the effects of natural hazard such as typhoons, floods, earthquakes, tsunamis, volcanic eruptions, and wildfires can be assessed, reducing damage and boosting national resilience. Satellites are also reliable and rapid tools for communication and navigation services that are crucial for search and rescue operations when ground-based infrastructure is disrupted. The use of space solutions for security activities is yet a process that can still be improved, in particular through the integration of space solutions with the data provided by terrestrial systems.

Within our measurement model, the "security" area has been assessed through the following indicators:

Indicator 54: National Security Policies. This indicator aims to assess the degree to which the use of space is integrated in the national security policies of the country and the degree to which the security of space assets is integrated in the national security policies of the country. Scores have been attributed by country experts invited to answer the following two questions (averaged to form a single score): "on a scale from 1 to 4, how well integrated would you say that space is in the national security policies of the country?"; and "on a scale from 1 to 4, how well integrated would you say that the security of space assets is in the national security policies of the country?" For both questions, scores range from 1 (not integrated at all) to 4 (fully integrated). A minimum of 3 responses have been submitted for each country. Answers from multiple respondents have been averaged to form a single score.

Indicator 55: Surveillance and Monitoring. This indicator aims to assess the degree to which the considered actor makes use of space assets for surveillance, verification, and/or risk assessment purposes. Examples include the use of space assets surveillance of illegal activities, compliance with international treaties verification, natural and human-made risk assessment, border control and maritime domain awareness. Scores have been attributed by country experts invited to answer the following survey question: "on a scale from 1 to 4, how often would you say that space assets are used for surveillance, verification, and/or risk assessment in the country?" Scores range from 1 (never) to 4 (very often). A minimum of 3 responses have been submitted for each country. Answers from multiple respondents have been averaged to form a single score.

Indicator 56: Disaster Prevention and Management. This indicator aims to assess the degree to which the considered actor makes use of space assets for crisis prevention and disaster management, both natural and human made. Hurricanes and storm forecasting, infrastructure decay monitoring, landslide early warnings, fire detection, support to search and rescue operations, intervention planning are few of the security applications that can be powered by space technologies for disaster prevention and management. Scores have been attributed by country experts invited to answer the following survey question: "on a scale from 1 to 4, how often would you say that space assets are used in crisis and disaster prevention and/or management in the country?" Scores range from 1 (never) to 4 (very often). A minimum of 3 responses have been submitted for each country. Answers from multiple respondents have been averaged to form a single score.

Defence

Space is a decisive tool for defence purposes and the conduct of military operations. The First Gulf War of 1991, also known as the "first space war," demonstrated the key added value of space in times of conflict. Subsequent armed interventions confirmed this and the recent war in Ukraine that started in February 2022 illustrated the key role that even commercial space actors can play to support military operations.

In a conflict, space solutions can be used a variety of applications (Bataille & Messina, 2020; DIA, 2022; JCS, 2020). Thus, remote sensing capacities (e.g., remote sensing, signal intelligence) provide intelligence, surveillance and reconnaissance that can help the warfighter, but also early warning against adversary missiles or meteorological information. Satellite communications allow the command and control of troops on the field, but also the control of drones and the reception of their data. Finally, PNT systems provide precise and accurate geo-location, navigation, and time reference services. PNT information is considered mission-essential for virtually every modern weapons system (JCS, 2020). In short, space is a ubiquitous infrastructure that an established space actor can leverage for its military actions.

The use of satellites for such vital functions has also led to increased threats towards these same space systems, making space a "contested" and even, for some countries, a "warfighting" domain. To mitigate these threats, different measures have been taken to make space systems and architectures more resilient. In addition, the importance of space surveillance has grown, moving towards Space Domain Awareness (which implies that the mission of the spacecraft is also known). Space surveillance also has an effect on the field. Indeed, it allows military commanders to know when their apparatus can be observed by the enemy, and to adapt their operations accordingly.

Within our measurement model, the "defence" area has been assessed through the following indicators:

Indicator 57: National Military Strategy. This indicator aims to assess the degree to which space is integrated in the military strategies and doctrines of the considered actor. The indicator in particular aims to assess whether military strategies consider the need to integrate satellites in conventional warfare, ensure their security and defence, while working to ensuring redundancy or lowering reliance on satellite technologies at the same time. It also considers whether a military strategy evaluates how to deny or limit access to space and space applications during conflicts. Scores have been attributed by country experts invited to answer the following survey: "on a scale from 1 to 4, how well integrated would you say that space is in the national military strategy of the country?" Scores range from 1 (not integrated at all) to 4 (fully integrated). A minimum of 3 responses have been submitted for each country. Answers from multiple respondents have been averaged to form a single score.

Indicator 58: Prevention and Deterrence. This indicator aims to assess the degree to what extent the considered actor makes use of space assets for prevention and deterrence purposes through e.g., space-based ISR, environmental monitoring (both terrestrial and space), missile early warning, nuclear denotation detection Scores have been attributed by country experts invited to answer the following survey question: "on a scale from 1 to 4, how often would you say that the country uses space for the prevention and/or deterrence of hostile actions?" Scores range from 1 (never) to 4 (very often). A minimum of 3 responses have been submitted for each country. Answers from multiple respondents have been averaged to form a single score.

Indicator 59: Military Operations. This indicator aims to assess the degree to which space assets are used for military operations in terms of command, control, communications, and computing (C4), intelligence, surveillance, and reconnaissance (ISR) as well as general military support services such as logistical support and missile guidance. Scores have been attributed by country experts invited to answer the following three questions (averaged to form a single score): (a) "on a scale from 1 to 4, how often would you say that the country uses space in military operations when it comes to command, control, communications, computing (C4)?"; (b) on a scale from 1 to 4, how often would you say that the country uses space in military operations when it comes to military intelligence, surveillance, and reconnaissance capabilities (ISR), including early warning, signal interception, and active observation?"; (c) on a scale from 1 to 4, how often would you say that the country uses space in military operations when it comes to other military support services (e.g., augmentation of terrestrial technologies as weather forecasting, data transfer, logistical support, missile guidance, etc.)?"

For each question, scores range from 1 (never) to 4 (very often). A minimum of 3 responses have been submitted for each country. Answers from multiple respondents have been averaged to form a single score.

Foreign Policy

The establishment of a world-class space programme can contribute to foreign policy matters in many different ways.

First, space is a field where countries can assert themselves globally by bolstering their image in the eyes of the citizens and the international community as a whole. By offering a clear evidence of national industrial, scientific and technological prowess, space achievements act as an important "status marker" that put countries at the forefront in the international arena and contributes to their image as a major player with the capacity to execute costly, complex and highly ambitious undertakings (Aliberti, 2015).

Second, the space programmes provide countries with an important instrument for its power projection on the international stage—functioning as both a hard and soft power tool. This function enables both the maintenance and expansion of hard power on the one hand but also the exercise of political leadership on important global issues such as, for instance, the fight against climate change, sustainable development and humanitarian aid.[2] From this perspective, a state-of-the-art space programme can be an enabler of international cooperation and contribute to making a country an indispensable partner for supporting global initiatives in key domains. Among these, particularly noteworthy are the contributions offered by space in the context of the implementation of the UN Sustainable Development Goals (SDGs). The combination of different satellite application systems like Earth Observation and PNT can indeed prove key to both monitoring the status of implementation and supporting the achievement of specific SDGs, including, among others, SDG-2 (zero hunger—for instance, in terms of crop productivity optimisation), SDG-6 (clean water and sanitation, for instance, in terms of water quality monitoring), SDG-7 (affordable and clean energy, for instance, in terms of power grid synchronisation), and SDG-13 (climate action, for instance in terms of climate change monitoring and disaster management) (UNOOSA, 2018).

Third, space can be used as a diplomatic toolkit to achieve broader national objectives, including political, strategic (i.e., military), and economic objectives. Inserting or reinforcing key space issue-areas into existing bilateral consultations and decision-making with third countries has often provided the two sides with an important tool to achieve common goals, e.g., deepen political ties and enhance economic exchanges (Robinson, 2011). Specifically, space cooperation can be a very symbolic—yet tangible—tool in a broader initiative aimed at building more effective strategic partnerships, as well as furthering the global impact of these partnerships on the international system. While space is only one of the issue areas where different countries can enhance cooperation, it can nonetheless serve as a catalyst for achieving a variety of shared policy objectives. Indeed, thanks to its pervasive utility and cross-sectorial impact, space provides a wide range of opportunities for the establishment of more tangible cooperation schemes with respect to such security issues as terrorism and maritime domain awareness as well as socio-economic issues such climate change, urban development, energy, environment, transport and sustainable development, to name a few (Aliberti, 2018).

Fourth, international space cooperation can, in turn, contribute to the optimal implementation of national space programmes, support the advancement of space research, innovation and state-of-the-art space and can also serve as a market-opener for the promotion of national technology and services outside the domestic markets,

[2] Space can for instance support capacity building in developing countries or assist populations, countries and regions facing natural or man-made disasters.

with great benefits for the national space industry. Finally, space cooperation can support the promotion of national values, interests and preferences on the international stage, including in terms of norms- and rule-setting in international fora dedicated to space.

Consistent with the above, within our measurement model, the "foreign policy" area has been assessed through the following four indicators:

Indicator 60: Space for Diplomacy. This indicator aims to assess the degree to which the considered actor makes uses of its space hard capacities to serve diplomatic purposes, be them political, strategic, and/or economic in nature. Scores have been attributed by country experts invited to answer the survey question: "on a scale from 1 to 4, how often would you say that the country uses space for diplomatic purposes, be them political, strategic, and/or economic?" Scores range from 1 (never) to 4 (very often). A minimum of 3 responses have been submitted for each country. Answers from multiple respondents have been averaged to form a single score.

Indicator 61: Diplomacy of Space. This indicator aims to assess the degree to which the considered actor is capable of exerting influence in norms diffusion and rulemaking process of major international decision-making fora related outer space (e.g., the United Nations Committee on the Peaceful Uses of Outer Space (COPUOS), the Conference on Disarmament (CD), the International Standardisation Organisation (ISO), the International Telecommunication Union (ITU), etc.). Scores have been attributed by country experts invited to answer the survey question: "on a scale from 1 to 4, how influential would you say that the country is in international space fora (e.g., COPUOS, CD, etc.)?" Scores range from 1 (not influential at all) to 4 (very influential). A minimum of 3 responses have been submitted for each country. Answers from multiple respondents have been averaged to form a single score.

Indicator 62: Foreign Aid and International Initiatives. This indicator aims to assess the degree to which the considered actor makes uses of its hard capacities to provide assistance to foreign countries without know-how or material capacities and contribute to major international initiatives like International Charter: Space and Major Disasters Charter or the UN-led Space4SDGs initiative. Scores have been attributed by country experts invited to answer the survey question: "on a scale from 1 to 4, how often would you say that the country uses space to support foreign aid and/or contribute to major international initiatives (e.g., UN SDGs)?" Scores range from 1 (never) to 4 (very often). A minimum of 3 responses have been submitted for each country. Answers from multiple respondents have been averaged to form a single score.

Indicator 63: International Prestige. This indicator aims to assess the degree to which the considered actor is successful in leveraging its achievements in space to boost its international reputation and contribute to its image as a major player in the international arena. Scores have been attributed by country experts invited to answer the survey question: "on a scale from 1 to 4, how successful would you say that the country is in using space to create/boost its international prestige?" Scores range from 1 (not successful at all) to 4 (very successful). A minimum of 3 responses have been submitted for each country. Answers from multiple respondents have been averaged to form a single score.

3.3 Measuring Autonomy

The autonomy dimension provides a second set of criteria for pursuing the comparative assessment of space actors. Autonomy refers to the state's ability to independently set directions, define policies, and operate in the space realm by relying on its own strengths and resources. Autonomy comprises two sub-dimensions, a technical and a political one. Accordingly, autonomy has been measured through two sub-indices.

- The *technical (or hard) autonomy sub-index*, based on 13 indicators, which aims to assess to which extent a state has the means to access and operate in space without relying on external sources of supply.
- The *political (or soft) autonomy sub-index*, based on 24 indicators, which aims to evaluate the state's ability to formulate interests of its own, independent of or against the will of divergent political and societal interests emanating from both inside and outside the country.

Both sub-indices comprise various areas, each encompassing several entries. A score ranging from 1 ("lowest autonomy") to 4 ("highest autonomy") has been attributed to each entry. Entries score have been consequently averaged for the corresponding area, ultimately leading to the compilation of a synthetic indicator for the "hard autonomy" sub-dimension and a synthetic indicator for the "soft capacity" sub-dimension.

Similar to the approach presented for capacity, the values of the two autonomy indices are transposed in the autonomy sub-matrix, which combines an actor's hard (technical) and soft (political) autonomy and represents its relative position in the sphere of autonomy (see Fig. 3.3).

As for capacity, eventually, the two autonomy sub-indices have been averaged and aggregated in a total autonomy index. This total autonomy value is used to plot the main matrix of space power.

3.3.1 *Hard (Technical) Autonomy*

With respect to technical autonomy, entries stretch across three prominent macro-areas: "production" (i.e., the technologies required for the development and manufacturing of space assets, including components, materials, and processes), "operations," and "exploitation." Table 3.3 features the indicators which have been used in our empirical assessment.

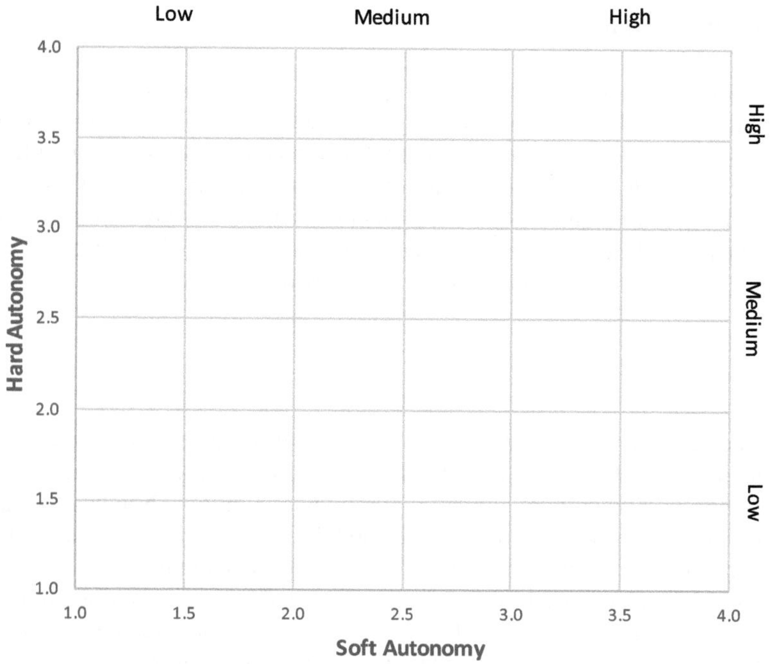

Fig. 3.3 Autonomy sub-matrix overview

Table 3.3 Measuring technical autonomy: macro-areas, areas, and entries

Macro-area	Area	Entry
Production phase	Space hardware	64: Launch vehicles
		65: Satellite systems
		66: Modules, capsules, and probes
	Infrastructure	67: Ground segment
		68: Assembly, integration, and testing facilities
Operations phase	Launch operations	69: Launch sites
		70: Orbital access
	Satellite operations	71: Space situational awareness
		72: Telemetry, tracking, and command
	Crewed operations	73: Astronaut training
		74: Human-rated operations
Exploitation phase	Data acquisition	75: Autonomous access to satellite data
	Service provision	76: Autonomous provision of satellite data-related services

3.3.1.1 Production Phase

The "production phase" macro-area can be broken down in two main areas and types of activities, namely: the manufacturing of space hardware (e.g., launch vehicles, satellite systems, payloads, capsules, modules), which represents the core infrastructure, and the build-up of the enabling infrastructure, which encompasses elements such as the ground segment for TT&C functions and the assembly, integration and testing facilities. Both are essential components for an autonomous access to and use of space. A description of these two areas is provided hereby.

Space Hardware

For an actor to be considered fully autonomous, it needs to possess the ability to indigenously manufacture its space hardware, which includes launch vehicles, satellite systems and subsystems, modules, capsules and probes. Needless to say, the production of all these elements should also take place within national borders, so that the country can decide on their use in self-determined manner.

Among these elements, the domestic production of launch vehicle arguably represents the most fundamental enabler for the conduct of autonomous space activities. Without indigenous launch capabilities, no country can really consider itself autonomous. In fact, the lack of autonomous access to space would cause dependency on foreign countries for space transportation needs, and thus restrict a country's political sovereignty vis-à-vis its ability to decide when and under what conditions to launch. In this respect, a major inherent risk is that a foreign launch provider could impose unacceptable conditions for launching satellites or even refuse to launch on the basis of political considerations (Aliberti & Tugnoli, 2015). Possessing a launch vehicle, however, is not enough, if its subsystems or critical components are not indigenously developed—the Atlas V's reliance on Russian-made RD-180 engines being a clear example of reliance on third countries for the production of critical technologies.

As a matter of fact, it is important to highlight that the most crucial aspect when assessing situations of dependency is not reliance on final products, but rather on basic technologies, since the lower the level of technology on which a country or region is dependent, the greater the weakness of the country or region (Caito, 2015). Actual technological autonomy in the production phase of any space hardware would need to be verified down to the level of Electrical, Electronic and Electromechanical (EEE) components, although this goes far beyond the space sector and no uniform, comparable data is available.

Within our measurement model, the "space hardware" area has been assessed through the following three indicators:

Indicator 64: Launch Vehicles Production. In order for a country to be autonomous in space activities, it must not rely on critical foreign technology for the production of launch vehicles. The element considered while scoring this indicator is the sovereign production of (a) spaceflight-related boosters; (b) spaceflight-related avionics; (c) rocket engines. Scores for this entry have been attributed as follows:

4 = all the elements have been built by the country on its own during the considered period (2011–2020).
3 = two elements have been built by the country on its own during the considered period (2011–2020).
2 = one element has been built by the country on its own during the considered period (2011–2020).
1 = no elements have been built by the country on its own during the considered period (2011–2020).

Indicator 65: Satellite Systems Production. For a country to be autonomous in space activities, it is crucial that a significant part of its satellite systems are manufactured within the country, enabling the country to access data of any quality desired, any time, and without limiting conditions externally imposed. If one actor relied on foreign-manufactured satellites, said country would be basing its space activities on the willingness of third-party countries to not impose restrictions as technology export control. Scores for this entry have been attributed as follows:

4 = more than 90% of the satellite systems launched in the period 2011–2020 have been manufactured in the country of ownership.
3 = between 75 and 90% of the satellite systems launched in the period 2011–2020 have been manufactured in the country of ownership.
2 = between 50 and 75% of the satellite systems launched in the period 2011–2020 have been manufactured in the country of ownership.
1 = 50% or more of the satellite systems launched in the period 2011–2020 have been manufactured elsewhere.

Indicator 66. Modules, Capsules, and Probes. To achieve autonomy in space exploration activities, a state must avoid dependency on critical hardware manufactured outside its borders. This indicator considers whether countries autonomously manufactured (a) modules, (b) capsules, and (c) probes for space exploration launched in the period 2011–2020.[3] Scores for this entry have been attributed as follows:

4 = all the elements have been manufactured by the country on its own during the considered period (2011–2020).
3 = two elements have been manufactured by the country on its own during the considered period (2011–2020).
2 = one element has been manufactured by the country on its own during the considered period (2011–2020).

[3] It is to be acknowledged that this indicator currently matches that of hard capacity. As the world transitions into commercially driven space exploration activities, more countries will be able to simply acquire capacity by purchasing space exploration hardware and services (e.g.: United Arab Emirates partnering with the American Universities). This will likely create a decoupling of hard capacity and technical autonomy.

1 = no element has been manufactured by the country on its own during the considered period (2011–2020).

Enabling Infrastructures

Possessing all the infrastructural elements that are required for manufacturing, deploying and operating space hardware is a prerequisite for attaining full technical autonomy. These elements include the Assembly, Integration and Testing facilities as well as the ground segment, which consists of all facilities and equipment that are used to monitor and control the assets in space.

Should an actor lack the necessary facilities to manufacture and integrate space hardware, it will not be able to autonomously build, launch, and operate spacecrafts.

Consistently, within our measurement model, the "enabling infrastructure" area has been assessed through the following two indicators:

Indicator 67: Ground Segment. In order for a country to be autonomous in space activities, it must be able to autonomously manufacture the major elements of a satellite ground segment so as ensure that it will be able to independently operate its satellite fleet. Elements considered within this entry are (a) Fixed Earth Stations for feeder links and TT&C; (b) Mobile Satellite terminals, and Very Small Aperture Terminals (VSATs). Scores for this entry have been attributed as follows:

4 = the country has built all the considered elements on its own.
3 = the country has built two elements on its own.
2 = the country has built one element on its own.
1 = the country has built no elements on its own.

Indicator 68: Assembly, Integration, and Testing Facilities. For a country to be autonomous, it needs to possess the necessary facilities to assemble, integrate and test its space hardware. This indicator considers institutional facilities located within national borders. Elements that have been considered critical for technical autonomy in this area are (a) clean rooms; (b) acoustic and vibration test chambers; (c) electromagnetic test chambers; thermal vacuum chambers. Scores for this entry have been attributed as follows:

4 = the country has access to institutional clean rooms and three types of additional facilities within national borders.
3 = the country has access to institutional clean rooms and two types of additional facilities within national borders.
2 = the country has access to institutional clean rooms and another type of additional facilities within national borders.
1 = The country has only access (or no access at all) to institutional clean rooms and three types of additional facilities within national borders.

3.3.1.2 Operations Phase

The "operations phase" macro-area has been segmented in three different areas that cover the main types of space operations, namely: launch operations, spacecraft and payload operations, and crewed operations. As for the production phase, these operations need to be undertaken in facilities within the country's national borders and/or under the direct control of its institutions. Mastering these three areas is considered an essential condition for ensuring a fully autonomous access to and use of space across the diverse types of actives. A description of these three areas is provided hereby.

Launch Operations

For an actor to be considered fully autonomous in launch operations, it should have access to launch site facilities that are located on its own territory and are also purpose-built. Equally important, it should have access to different classes of launchers that enable to reach the geostationary orbit (GTO), low and medium earth orbits (LEO, MEO) and deep space are even more concentrated, being mastered by only six space powers: the United States, Russia, Europe. Consistently, within our measurement model, the "launch operations" area has been assessed through the following three indicators:

Indicator 69: Launch Sites. For a country to be fully autonomous in accessing space, its launch site facilities have to be devoid of constraints and be fully fit-for-purposes. Countries should have a mix of launch sites that are: (a) within one own's territory; (b) in proximity of oceans or uninhabited areas; and (c) at ideal latitudes to enable access to multiple orbits. These three elements, together with data of the launch sites active in 2011–2020, have been considered when scoring this indicator. Scores for this indicator have been attributed as follows:

4 = the country's launch sites possess all the three elements.
3 = the country's launch sites possess two elements.
2 = the country's launch sites possess one element.
1 = the country's launch sites possess no elements (or no launch site at all).

Indicator 70: Orbital Access. In order for a country to be fully autonomous in reaching any desired orbit, it needs access to launchers of different classes. A country may have the capacity to launch and host multiple launch service providers, but it will not be autonomous if its programmes rely on launchers built abroad and operated by third countries' providers. This indicator divides the launchers in four classes: (a) small-lift; (b) medium-lift; (c) heavy-lift; and (d) superheavy-lift launch vehicles.

Scores are assigned based on active (i.e., effectively used) launchers in the period 2011–2020. Scores for this indicator have been attributed as follows:

4 = three or more classes of launchers are covered by national service providers.
3 = two classes of launchers are covered by national service providers.
2 = one class of launchers is covered by national service providers.
1 = no launcher.

Satellite Operations

For an actor to be considered fully autonomous in satellite operations (spacecraft and payload operations), it should have access to SSA data and services as well as TT&C functions. SSA is vital to operate in space—no qualified operator can afford to operate a satellite without proper access to SSA data and services (e.g., collision avoidance collision avoidance services, re-entry analysis, and fragmentation analysis). Equally important is that national TT&C systems are able to perform all the necessary tasks to ensure the successful operation of the country's satellite (e.g., monitoring of the health and status of the satellite determining the satellite's exact location, and controlling the satellite through the reception, processing, and implementation of commands transmitted from the ground (Guest, 2013).

Within our measurement model, the "satellite operations" area has been assessed through the following two indicators:

Indicator 71: Access to Space Situational Awareness Data and Services. In order for a country to autonomously operate in space, access to indigenously produced SSA data is necessary. This is to avoid having one own's operations depending on third-party will, including third-party filtering out information or quality of said data. Scores for this indicator have been attributed as follows:

4 = the country is fully autonomous in accessing SSA data and products in its space operations.
3 = the country is autonomous for most of its space operations.
2 = the country is reliant on third-party data and products for most of its space operations.
1 = the country is fully dependent on third-party data and products.

Indicator 72: Access to Telemetry, Tracking and Command Functions. In order for a country to autonomously operate in space, it needs to be able execute all necessary TT&C functions from nationally controlled facilities, without the necessity to rely on ground stations abroad. Nowadays space operations have become inherently global in nature, with operators and governments owning ground stations worldwide thanks to bilateral and multilateral agreements. This indicator considers whether the country would be able to autonomously operate satellites, and at what level of operational limitations due to geography—given that countries of different sizes, and at different latitudes, would have different advantages and disadvantages in tracking and communicating with satellites. Scores for this indicator have been attributed as follows:

4 = the country is autonomous in tracking and commanding satellites, from its own territory and globally through other means as data relay satellites and tracking ships.
3 = the country is autonomous in tracking and commanding satellites from its own territory without operational impairments.
2 = the country is autonomous in tracking and commanding satellites from its own territory with operational limitations.
1 = the country is not autonomous in tracking and commanding its space missions.

Crewed Operations

Attaining autonomy in crewed operations is a resource consuming effort that requires large amounts of financial investment as well as relevant experience and major technical skills. Countries aiming to conduct human spaceflight activities require a trained astronaut corps, which is a capital-intensive task from both a human and financial perspective. Smaller countries have been relying on international partnerships to train and launch their astronauts, but at the cost of giving up the ability to decide what the astronauts will do once in orbit. Countries aiming to conduct autonomous human spaceflight activities also need to have the ability to indigenously launch and recover astronauts through human-rated launchers and crewed capsules.

Within our measurement model, the "launch operations" area has been assessed through the following three indicators:

Indicator 73: Astronaut Training. In order for a country to be autonomous in crewed operations, it needs to be able to train its astronauts, without the need to rely on third countries' training facilities. The presence of training facilities can be used as a proxy for the level of autonomy a country can have in training its astronaut corp. Scores for this indicator have been attributed as follows:

4 = the country has advanced facilities, and does not need to send abroad its astronauts for training.
3 = the country has advanced facilities, still relying on foreign facilities for some training.
2 = the country has basic facilities, but its astronaut's corps requires major training abroad.
1 = the country relies almost or entirely on foreign facilities.

Indicator 74: Human-Rated Operations. In order for a country to be autonomous in crewed operations, it needs to have the ability to indigenously launch and recover its astronauts. To do so, a country requires: (a) human-rated launchers; (b) crewed capsules; (c) in-space human spaceflight experience.[4] Scores for this indicator have been attributed as follows:

4 = the country has all the elements indicated.
3 = the country has two of the elements.
2 = the country has one of the elements.
1 = the country has no experience nor launchers.

3.3.1.3 Exploitation Phase

The "exploitation phase" macro-area has been segmented in two different areas that cover the main types of activities that need to be performed to effectively exploit satellite systems once in orbit, namely: data acquisition and service provision. Mastering these two areas is considered an essential condition for ensuring a truly autonomous use of space across the diverse types of actives. A description of these three areas is provided hereby.

Data Acquisition

Autonomous access to satellite data is an important condition for ensuring a proper exploitation of space assets. This is only possible when countries have nationally manufactured and owned assets across the different types of applications (e.g., telecommunications satellites, remote sensing, science, SSA, PNT satellites). Within our measurement model, the "data acquisition" area has been assessed through the following indicator:

Indicator 75: Autonomous Access to Satellite Data. To achieve autonomy in the exploitation of space-related data, a country must possess independent access to such data. This indicator considers whether one country has autonomous access to: (a) satcom data; (b) Earth Observation data; (c) Position, Navigation, and Timing data; and (d) Space Situational Awareness data. The indicator considers the spacecraft launched in the period 2011–2020, and whether countries have access to data generated through sovereign capabilities (i.e., nationally manufactured and owned capabilities). Scores for this indicator have been attributed as follows:

4 = all the applications covered by sovereign data/spacecraft.
3 = three satellite applications covered by sovereign data/spacecraft.
2 = two satellite applications covered by sovereign data/spacecraft.
1 = one or no applications covered by sovereign data/spacecraft.

[4] It is to be acknowledged that this indicator currently mirrors that of Hard Capacity. As the world transitions towards more commercially-driven human spaceflight activities, more countries will be able to simply acquire capacity by purchasing human spaceflight hardware and services. This will create a decoupling of Hard Capacity and Technical autonomy.

Service Provision

Once data access has been secured, products and services need to be developed and provided. A space service can be defined as an accessible application derived from a space-based technological system and the subsequent data and provisions it provides with facilitated use for the end-user. In this sense, a service is a product of (space-related) research and innovation, going beyond the provision of data, to having a functioning application for a defined user base. Services can be funded by and conducted via varying public, private, and public and private models within a considered polity. Service provision to end-users (be them governmental, commercial or individual users) by national entities is therefore another important condition for ensuring a proper exploitation of space assets.

Within our measurement model, the "service provision" area has been assessed through the following indicator:

Indicator 76: Autonomous Provision of Satellite Data-Related Services. In order for a country to be autonomous in exploiting space-related data, it must have autonomous capabilities to provide services related to said data. This indicator considers whether one country has national actors delivering services related to (a) satcom data; (b) remote sensing data; (c) Position, Navigation, and Timing data; and (d) Space Situational Awareness data.[5] Scores for this indicator have been attributed as follows:

4 = all the applications covered by national actors.
3 = three satellite applications covered by national actors.
2 = two satellite applications covered by national actors.
1 = one or no applications covered by national actors

3.3.2 *Soft (Political) Autonomy*

Measuring political autonomy requires assessing the state sovereignty over space affairs. As thoroughly analysed by S. Krasner, the term sovereignty can have four different meanings: (a) *international legal sovereignty*, which "refers to the practices associated with mutual recognition, usually between territorial entities that have formal juridical independence"; (b) Westphalian sovereignty, which refers to political organisation based on the exclusion of external actors from authority structures within a given territory"; *domestic sovereignty*, which "refers to the formal organisation of political authority within the state and the ability of public authorities to

[5] It is worth acknowledging that once access to data—be them in-house generated or sourced from third-party actors—is secured, it is relatively straightforward to ensure service provision. Almost any country has governmental agencies, academic institutions, or private companies that can deliver space-based data products and services. The quality of such product should be measured together with soft capacity indicators.

exercise effective control within the borders of their own polity"; and *interdependence sovereignty*, which "refers to the ability of public authorities to regulate the flow of information, ideas, goods, people, pollutants, or capital across the borders of their state" (Krasner, 1999: 3–4).

Within this book, we concentrate in particular on Westphalian sovereignty and domestic sovereignty and posit that soft (political) autonomy concerns the state's ability to formulate space-related policies that it has autonomously determined irrespective of conflicting particular interests emanating inside or outside its national borders. As already noted in Sect. 2.2 in Chap. 2, however, it is important to highlight that autonomy should be understood as a matter of degree, and as such, it should not be forced into rigid typological classifications. While ideal–typical definitions can be used for conceptual clarity—as we ourselves do—they should be used in relative rather than absolute formulations. Few polities in the world are totally "stateless" and completely dominated by business organisations. Similarly, few, if any polities in the world are totally dominated by the state—this was not even the case in the USSR and China under conditions of full-fledged nationalisation. In addition, the degree of autonomy may vary according to the policy domain under consideration; evaluating and measuring it may be feasible, empirically, only by keeping this aspect in mind. Different policies, in fact, mobilise actors of different nature, weight and resources, which give rise to different dynamics of influence and balance of power. In any country, the state, depending on multiple circumstances, can prove more resilient to any agent of influence in one policy area and less resilient in another.

Regarding the agents of influence with the potential to challenge state autonomy over space matters, several types can be identified. These include *external agents* such as foreign states, multinational corporations, and international organisations as well as *internal agents* such as the military and domestic corporations, both private and state-owned. For the sake of simplicity, the analysis of the agents of influence has been limited to foreign states, the national military and domestic corporations.

Within our measurement model soft (political) autonomy is measured across two fundamental macro-areas: "external decisions" and "internal decisions." Both macro-areas have been broken down in three areas each. Entries for measuring political autonomy in each of these areas have been summarised in Table 3.4.

Values for the "political autonomy" index result from the aggregation of 18 expert-coded indicators. Scores for each indicator composing the index have been attributed by respondents who participated in our ad hoc "measuring spacepower" survey.

3.3.2.1 External Decisions

The "external decisions" macro-area aims to assess the degree to which an actor is autonomous from different agents of influence (both inside and outside the polity) when it comes to taking decisions about the political behaviour to be adopted in the international arena. In particular, we concentrate on the autonomy of the actor in: (a) deciding whether or not to enter into an international agreement or treaty and/or to become a member of an international organisation or forum; (b) choosing a political

Table 3.4 Measuring political autonomy: macro-areas, areas, and entries

Macro-area	Area	Entry
External decisions	Joining space-related agreements and organisations	77: From foreign nations
		78: From the military
		79: From domestic corporations
	Acting within space-related multilateral organisations	80: From foreign nations
		81: From the military
		82: From domestic corporations
	Complying with space-related international arrangements	83: From foreign nations
		84: From the military
		85: From domestic corporations
Internal decisions	Formulating a national space policy	86: From foreign nations
		87: From the military
		88: From domestic corporations
	Defining a national space programme	89: From foreign nations
		90: From the military
		91: From domestic corporations
	Choosing partners within the space domain	92: From foreign nations
		93: From the military
		94: From domestic corporations

course of action (e.g. voting, vetoing and coalition building) within multilateral forums or organisations of which the State is a member; (c) determining whether to comply with international obligations to which a state has formally committed itself, whether these are legally binding tools, such as treaties, or soft-law instruments such as guidelines, best practices, rules of conduct, or technical standards.

Consistent with this categorisation, the "external decisions" macro-area covers three main types of decisions, dubbed as: joining space-related treaties and fora, acting within space-related multilateral fora, complying with space-related international arrangements.

There are multiple international *fora* in which principles, norms, rules, standards and procedures relevant to the conduct of space activities are being promoted, discussed and agreed upon. The most relevant during the timeframe considered in this study (2011–2020) are:

- The United Nations General Assembly, the Fourth Committee and the First Committee.
- The United Nations Committee on the Peaceful Uses of Outer Space (COPUOS) and its two sub-committees.
- The Conference on Disarmament (CD).
- The 2012/13 Groups of Governmental Experts on Transparency and Confidence Building Measures (TCBMs).

- The 2018/19 Groups of Governmental Experts on the Prevention of an Arms Race in Outer Space (PAROS).
- The Open-Ended Working Group on Reducing Space Threats (OEWG).
- The Inter-Agency Space Debris Coordination Committee (IADC).
- The International Committee on GNSS (ICG).
- The Intergovernmental Group on Earth Observation.
- The Committee on Earth Observation Satellites.
- The International Organisation for Standardisation (ISO).
- The Consultative Committee for Space Data Systems (CCSDS).
- The International Telecommunication Union (ITU).

As for the international arrangements considered by the study, these include both legally binding tools and soft-law instruments such as:

- The main international treaties related to outer space (specifically the 1963 "Partial Test Ban Treaty," the 1967 "Outer Space Treaty", the 1968 "Rescue Agreement", the 1972 "Liability Convention", the 1975 Convention on Registration of Objects Launched into Outer Space (the "Registration Convention", the 1979 "Moon Treaty").
- The UN Five Declarations and Legal Principles (specifically, the 1963 "Declaration of Legal Principles," the 1982 "Broadcasting Principles," the 1986 "Remote Sensing Principles," the 1992 "Nuclear Power Sources Principles," the 1996 "Benefits Declaration").
- The annual resolutions of the United Nations General Assembly.
- The IADC Space Debris Mitigation Guidelines.
- The Guidelines for the Long-term Sustainability of Space Activities.
- The Transparency and Confidence Building Measures established by the GGE 2012/13.
- The Draft International Code of Conduct for Outer Space Activities (ICoC).
- The Draft Treaty Draft Treaty on the Prevention of the Placement of Weapons in Outer Space, the Threat or Use of Force against Outer Space Objects (PPWT).
- The Artemis Accords.
- ISO standards.
- Radio Regulations of the World Radiocommunication Conferences (WRC).

Joining Space-Related Agreements and Organisations

Our main concern here is assessing the autonomy of the actor in deciding whether or not to enter into an international agreement or treaty and/or to become a member of an international organisation or forum. Each indicator is devoted to assessing such autonomy from a specific agent of influence, namely foreign nations, military actors and corporations (both public and private). Within our measurement model, the "joining space-related agreements and organisations" area has been more specifically assessed through the following indicators:

Indicator 77: Autonomy from Foreign Nations in "Joining." This indicator assesses the level of autonomy that State possesses in deciding to sign an international space-related arrangement (be it legally binding or not) and/or to become member of a space-related international organisation or forum, independently from foreign nations. This indicator focuses on the State's ability to independently pursue its will in the international space arena, regardless of its contingent definition or content, by deciding to join or not to join international agreements and fora: in doing so, it does not discriminate among states on the basis of the number of treaties signed or organisations joined and does not assume "isolationist" powers to be by definition more autonomous than "internationalist" ones. Specifically, the indicator allows for evaluating the robustness of the state to foreign invitations and interventions in the treaty-making and membership application processes. Scores have been attributed by country experts invited to answer the survey question: "on a scale from 1 to 4, how autonomous from foreign nations would you say that the country is when it comes to the decision to join space-related bilateral and/or multilateral arrangements? Scores range from 1 (not autonomous at all) to 4 (fully autonomous). A minimum of 3 responses have been submitted for each country. Answers from multiple respondents have been averaged to form a single score.

Indicator 78: Autonomy from the Military in "Joining." This indicator assesses the level of autonomy that a State possesses in deciding to sign an international space-related agreement (be it legally binding or not) and/or to become member of a space-related international organisation or forum, independently from the military. Scores have been attributed by country experts invited to answer the survey question: "on a scale from 1 to 4, how autonomous from the national military would you say that the country is when it comes to the decision to join space-related bilateral and/or multilateral arrangements?" Scores range from 1 (not autonomous at all) to 4 (fully autonomous). A minimum of 3 responses have been submitted for each country. Answers from multiple respondents have been averaged to form a single score.

Indicator 79: Autonomy from Corporations in "Joining." This indicator gauges to what extent the State maintains autonomy from domestic corporations (both public and private) when it comes to deciding to sign an international space-related agreement (be it legally binding or not) and/or to become member of a space-related international organisation or forum. Scores have been attributed by country experts invited to answer the survey question: "on a scale from 1 to 4, how autonomous from domestic corporations (state or private) would you say that the country is when it comes to the decision to join space-related bilateral and/or multilateral arrangements?" Scores range from 1 (not autonomous at all) to 4 (fully autonomous). A minimum of 3 responses have been submitted for each country. Answers from multiple respondents have been averaged to form a single score.

Acting Within Space-Related Multilateral Organisations

Our goal here is to assess the autonomy of the actor in choosing a political course of action (e.g., voting, vetoing and coalition building) within multilateral fora or organisations of which it is a member. Each indicator is devoted to assessing such autonomy from a specific agent of influence, namely foreign nations, military actors and corporations (both public and private). Within our measurement model, the "acting within space-related multilateral organisations" area has been more specifically assessed through the following three indicators:

Indicator 80: Autonomy from Foreign Nations in "Acting." This indicator assesses the degree to which the State is autonomous from foreign nations when it comes to acting within space-related multilateral organisations and fora considered above. This indicator focuses on the actor's ability to independently formulate its interests in the international arena, regardless of their contingent definition or content, by participating to the activities of space-related organisations. Specifically, the indicator allows for evaluating the robustness of the state to foreign invitations and interventions with regard to activities such as voting, vetoing, and coalition-building within international fora. Scores have been attributed by country experts invited to answer the survey question: "on a scale from 1 to 4, how autonomous from foreign nations would you say that the country is when it comes to acting (e.g., voting, coalition building, etc.) within major international space fora?" Scores range from 1 (not autonomous at all) to 4 (fully autonomous). A minimum of 3 responses have been submitted for each country. Answers from multiple respondents have been averaged to form a single score.

Indicator 81: Autonomy from the Military in "Acting" This indicator assesses the degree to which the State is autonomous from the domestic military when it comes to acting within space-related multilateral organisations and fora. Scores have been attributed by country experts invited to answer the survey question: "on a scale from 1 to 4, how autonomous from the national military would you say that the country is when it comes to acting (e.g., voting, coalition building, etc.) within major international space fora?" Scores range from 1 (not autonomous at all) to 4 (fully autonomous). A minimum of 3 responses have been submitted for each country. Answers from multiple respondents have been averaged to form a single score.

Indicator 82: Autonomy from Domestic Corporations "In Acting." This indicator assesses the degree to which the State is autonomous from domestic corporations (both public and private) when it comes to acting within space-related multilateral organisations and fora. Scores have been attributed by country experts invited to answer the survey question: "on a scale from 1 to 4, how autonomous from domestic corporations (state or private) would you say that the country is when it comes to acting (e.g., voting, coalition building, etc.) within major international space fora?" Scores range from 1 (not autonomous at all) to 4 (fully autonomous). A minimum of 3 responses have been submitted for each country. Answers from multiple respondents have been averaged to form a single score.

Complying With Space-Related International Law

This area aims to assess the autonomy of the actor in determining whether to comply with international obligations to which a state has formally committed itself, irrespective of whether these are derived from legally binding tools or soft-law instruments such as guidelines, best practices, rules of conduct, or technical standards. Each indicator is devoted to assessing such autonomy from a specific agent of influence, namely foreign nations, military actors and corporations (both public and private). The actor is considered autonomous when the decisions have not been determined by these agents of influence. Within our measurement model, the "acting within space-related multilateral organisations" area has been more specifically assessed through the following three indicators:

Indicator 83: Autonomy from Foreign Nations in "Complying." This indicator assesses the degree to which the state is autonomous from foreign nations when it comes to deciding to (or not to) comply with space-related international decisions (be them legally binding or soft-law in nature). This indicator focuses on the actor's ability to independently pursue its decision to commit or not to commit to international obligations, rather than on its sheer level of compliance with law and regulations. Specifically, the indicator allows for evaluating the robustness of the state to foreign invitations and interventions with regard to abiding by international law, be it "soft" or "hard," customary or treaty based. Scores have been attributed by country experts invited to answer the survey question: "on a scale from 1 to 4, how autonomous from foreign nations would you say that the country is when it comes to complying with space-related international law (including both 'hard' and 'soft' law)?" Scores range from 1 (not autonomous at all) to 4 (fully autonomous). Answers from multiple respondents have been averaged to form a single score.

Indicator 84: Autonomy from the Military "In Complying." This indicator assesses the degree to which the State is autonomous from the domestic military when it comes to deciding to comply with space-related international regulations. Scores have been attributed by country experts invited to answer the survey question: "on a scale from 1 to 4, how autonomous from the national military would you say that the country is when it comes to complying with space-related international law (including both 'hard' and 'soft' law)?" Scores range from 1 (not autonomous at all) to 4 (fully autonomous). Answers from multiple respondents have been averaged to form a single score.

Indicator 85: Autonomy from Domestic Corporations in "Complying." This indicator assesses the degree to which the State is autonomous from domestic corporations (both public and private) when it comes to deciding to comply with space-related international regulations. Scores have been attributed by country experts invited to answer the survey question: "on a scale from 1 to 4, how autonomous from domestic corporations (state or private) would you say that the country is when it comes to complying with space-related international law (including both 'hard' and 'soft' law)?" Scores range from 1 (not autonomous at all) to 4 (fully autonomous). Answers from multiple respondents have been averaged to form a single score.

3.3.2.2 Internal Decisions

The "internal decisions" macro-area aims to assess the degree to which an actor is autonomous from different agents of influence (both inside and outside its national borders) when it comes to taking decisions about the political behaviour to be adopted in the domestic context. In particular, we concentrate on the autonomy of the actor in: (a) formulating national strategies, policies, laws and regulations relevant to the conduct of space activities within the country; (b) designing operational programmes that outline how such policies are to be implemented by national and subnational agencies, public and private industry, research institutes etc.; (c) selecting partners for policy implementation and programme execution, which includes the allocation of resources and incentives to such entities (e.g., public contracts, public–private partnerships, domestic-foreign cooperation arrangements, etc.).

Consistent with this categorisation, the "internal decisions" macro-area covers three main types of decisions, dubbed as: formulating a national space policy, defining a national space programme, and choosing partners within the space domain. A description of these three areas is provided hereby.

Formulating a National Space Policy

This area assesses the decision-making autonomy of the state in formulating policies pertaining to the conduct of space activities within the country. Space policy is a broad umbrella term designating the deliberate course of action set by the state's highest executive branches with respect to the conduct of space activities by national entities. The term includes public policies in *stricto sensu,* strategic documents, guidelines, laws and regulations. This indicator focuses on the state's political ability to independently define its space policies, rather than on their content or scope.

Each indicator is devoted to assessing such autonomy from a specific agent of influence, namely foreign nations, military actors and corporations (both public and private). The actor is considered autonomous when the national space policies have not been shaped or determined by these agents of influence. More specifically, within our measurement model, the area "formulating a national space policy" has been more specifically assessed through the following three indicators:

Indicator 86: Autonomy from Foreign Nations in "Formulating Policies." This indicator assesses the degree to which the actor is autonomous from foreign nations when it comes to formulating its national space policy. Scores have been attributed by country experts invited to answer the survey question: "on a scale from 1 to 4, how autonomous from foreign nations would you say that the country is when it comes to formulating the national space policy?" Scores range from 1 (not autonomous at all) to 4 (fully autonomous). A minimum of 3 responses have been submitted for each country. Answers from multiple respondents have been averaged to form a single score.

Indicator 87: Autonomy from the Military in "Formulating Policies." This indicator assesses the degree to which the State is autonomous from the domestic military when it comes to formulating its national space policy. Scores have been attributed by country experts invited to answer the survey question: "on a scale from 1 to 4, how autonomous from the national military would you say that the country is when it comes to formulating the national space policy?" Scores range from 1 (not autonomous at all) to 4 (fully autonomous). A minimum of 3 responses have been submitted for each country. Answers from multiple respondents have been averaged to form a single score.

Indicator 88: From Domestic Corporations "In Formulating Policies." This indicator assesses the degree to which the State is autonomous from domestic corporations (both public and private) when it comes to formulating its national space policy. Scores have been attributed by country experts invited to answer the survey question: "on a scale from 1 to 4, how autonomous from domestic corporations (state or private) would you say that the country is when it comes to formulating the national space policy?" Scores range from 1 (not autonomous at all) to 4 (fully autonomous). A minimum of 3 responses have been submitted for each country. Answers from multiple respondents have been averaged to form a single score.

Defining a National Space Programme

Our objective here is to assess the decision-making autonomy of the state in defining its national space programmes. Space programmes are here indented as the operational activities carried out by national space agencies, public and private, and scientific institutions to implement the national space strategies and policies. The area more specifically focuses on the state's political ability to independently formulate its space programme (*decision-making process*), rather than on its management and execution (*decision-enforcement process*).

Each indicator is devoted to assessing such autonomy from a specific agent of influence, namely foreign nations, military actors and corporations (both public and private). The actor can be considered autonomous when the national space policies have not been shaped or determined by these agents of influence. More specifically, within our measurement model, the area "formulating a national space policy" has been more specifically assessed through the following three indicators:

Indicator 89: Autonomy from Foreign Nations in "Defining Programmes." This indicator assesses the degree to which the State is autonomous from foreign nations when it comes to defining its national space programme. Scores have been attributed by country experts invited to answer the survey question: "on a scale from 1 to 4, how autonomous from foreign nations would you say that the country is when it comes to defining a national space programme?" Scores range from 1 (not autonomous at all) to 4 (fully autonomous). A minimum of 3 responses have been submitted for each country. Answers from multiple respondents have been averaged to form a single score.

Indicator 90: Autonomy from the Military "in Defining Programmes." This indicator assesses the degree to which the State is autonomous from the domestic military when it comes to defining its national space programme. Scores have been attributed by country experts invited to answer the survey question: "on a scale from 1 to 4, how autonomous from the national military would you say that the country is when it comes to defining a national space programme?" Scores range from 1 (not autonomous at all) to 4 (fully autonomous). A minimum of 3 responses have been submitted for each country. Answers from multiple respondents have been averaged to form a single score.

Indicator 91: Autonomy from Domestic Corporations in "Defining Programmes." This indicator assesses the degree to which the State is autonomous from domestic corporations (both public and private) when it comes to defining its national space programme. Scores have been attributed by country experts invited to answer the survey question: "on a scale from 1 to 4, how autonomous from domestic corporations (state or private) would you say that the country is when it comes to defining a national space programme?" Scores range from 1 (not autonomous at all) to 4 (fully autonomous). A minimum of 3 responses have been submitted for each country. Answers from multiple respondents have been averaged to form a single score.

Choosing Partners Within the Space Domain

This area is concerned with assessing the decision-making autonomy of the state in selecting partners for the implementation of its policies and the execution of its programmes, especially in domains of strategical relevance such as technological supply and assistance. The area more specifically focuses on the state's ability to independently choosing not only specific cooperation partners (e.g., domestic entities, foreign states, foreign corporations, etc.) but also the specific cooperation schemes, including the allocation of resources and the incentives to cooperating entities (e.g., public contracts, public–private partnerships, domestic-foreign cooperation arrangements, etc.).

Each indicator is devoted to assessing such autonomy from a specific agent of influence, namely foreign nations, military actors and corporations (both public and private). The actor can be considered autonomous when the national space policies have not been shaped or determined by these agents of influence. More specifically, within our measurement model, the area "formulating a national space policy" has been more specifically assessed through the following three indicators:

Indicator 92: Autonomy from Foreign Nations in "Choosing Partners." This indicator assesses the degree to which the State is autonomous from foreign nations when it comes to choosing its partners within the space policy domain.

Scores have been attributed by country experts invited to answer the survey question: "on a scale from 1 to 4, how autonomous from foreign nations would you say that the country is when it comes to choosing its partners within the space domain?" Scores range from 1 (not autonomous at all) to 4 (fully autonomous). A minimum of 3 responses have been submitted for each country. Answers from multiple respondents have been averaged to form a single score.

Indicator 93: Autonomy from the Military in "Choosing Partners." This indicator assesses the degree to which the State is autonomous from the domestic military when it comes to choosing its partners within the space policy domain.

Scores have been attributed by country experts invited to answer the survey question: "on a scale from 1 to 4, how autonomous from the national military would you say that the country is when it comes to choosing its partners within the space domain?" Scores range from 1 (not autonomous at all) to 4 (fully autonomous). A minimum of 3 responses have been submitted for each country. Answers from multiple respondents have been averaged to form a single score.

Indicator 94: Autonomy from Domestic Corporations "In Choosing Partners." This indicator assesses the degree to which the State is autonomous from domestic corporations (both public and private) when it comes to choosing its partners within the space policy domain. Scores have been attributed by country experts invited to answer the survey question: "on a scale from 1 to 4, how autonomous from domestic corporations (state or private) would you say that the country is when it comes to choosing its partners within the space domain?" Scores range from 1 (not autonomous at all) to 4 (fully autonomous). A minimum of 3 responses have been submitted for each country. Answers from multiple respondents have been averaged to form a single score.

3.4 Indexing Capacity and Autonomy

The combination of the autonomy and capacity indices results in the so-called *spacepower matrix* (see Fig. 3.4).

The matrix allows for mapping space actors along two fundamental axes. It provides an ideal–typical representation of space actors' relative strengths based on the intersection of the capacity and autonomy scores: specifically, capacity scores determine the location of countries on the Y axis, while autonomy scores have been plotted on the X axis. For analytical purposes, the surface area delimited by the autonomy and capacity axes can be divided into four main cells or quadrants.

In Fig. 3.4 values are consistent with the fourfold scoring system adopted in this study. Demarcation lines enable to differentiate among states which obtain a low,

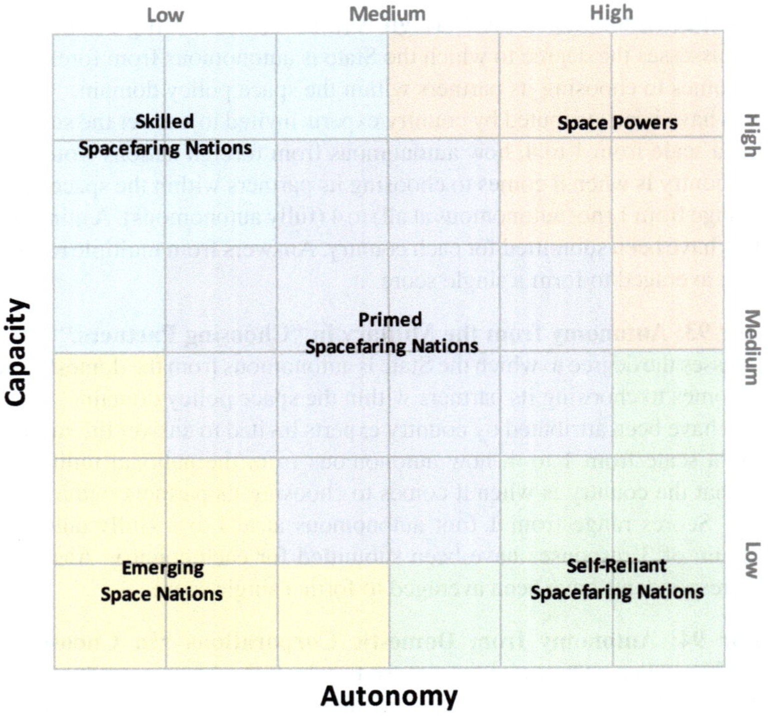

Fig. 3.4 Spacepower matrix overview

medium or high value on the capacity and autonomy scale. Most importantly, the typology of space actors presented in Fig. 3.4 builds on the scores reported on the capacity and the autonomy indices and the taxonomy defined in Chap. 2.

Also, the capacity and autonomy indices can be combined into a composite synthetic measure, the "spacepower index," which is a composite indicator comprising the two constituent dimensions of *spacepower*.

References

Aliberti, M. (2015). *When China Goes to the Moon*. Springer

Aliberti, M. (2018). *India in Space Between Utility and Geopolitics*. Springer.

Aliberti, M., & Tugnoli M. (2015). *European Launchers Between Commerce and Geopolitics* (ESPI Report 56). European Space Policy Institute.

Bataille, M., & Hrozensky, T. (2022). *Space in Support of Security Missions* (ESPI Report 80). European Space Policy Institute.

Bataille, M., & Messina, V. (2020). *Europe, Space and Defence* (ESPI Report 72). European Space Policy Institute.

Caito, L. (2015). *European Technological Non-Dependence in Space* (ESPI Report 51). European Space Policy Institute.

Cunningham, F. (2009). The Stellar Status Symbol: True Motives for China´s Manned Space Program. *China Security, 5*(3), 73–88.

DIA. (2022). *Challenges to Security in Space. Space Reliance in An Era of Competition and Expansion.* Defence Intelligence Agency. https://www.dia.mil/Portals/110/Documents/News/Military_Power_Publications/Challenges_Security_Space_2022.pdf (Accessed 10 February 2023)

DoD. (2022). *DoD Directive 3100.10 Space Policy*. United States Department of Defense—Office of the Under Secretary of Defense for Policy. https://www.esd.whs.mil/Portals/54/Documents/DD/issuances/dodd/310010p.PDF (Accessed 15 December 2022),

ESA. (2009). *What is Space Weather. European Space Agency*. https://swe.ssa.esa.int/what-is-space-weather (Accessed 22 January 2023).

ESA. (2023a). *Telecommunications Satellites*. European Space Agency. https://www.esa.int/Applications/Telecommunications_Integrated_Applications/Telecommunications_satellites (Accessed 10 January 2023).

ESA. (2023b). *About asteroids and Planetary Defence*. European Space Agency. https://www.esa.int/Space_Safety/About_asteroids_and_Planetary_Defence (Accessed on 20 January 2023).

ESA. (2023c). *About Space Science*. European Space Agency.

ESA. (2023d). *ESA's Climate Change Initiative*. European Space Agency. <https://www.esa.int/Applications/Observing_the_Earth/Space_for_our_climate/ESA_s_Climate_Change_Initiative> (Accessed 1 February 2023).

EUMETSAT. (2023). *What We Do*. European Organisation for the Exploitation of Meteorological Satellites. https://www.eumetsat.int/about-us/what-we-do (Accessed 5 December 2022)

EUSatCen. (2023). SSA. EUSatCen. https://www.satcen.europa.eu/page/ssa (Accessed 10 May 2023)

EUSPA. (2021a). *What is Earth Observation?* European Union Agency for the Space Programme. https://www.euspa.europa.eu/european-space/eu-space-programme/what-earth-observation (Accessed 1 February 2023).

EUSPA. (2021b). *What is GNSS?* European Union Agency for the Space Programme. https://www.euspa.europa.eu/european-space/eu-space-programme/what-gnss (Accessed 1 February 2023).

Gasparini, G., & Miranda, V. (2010). *Space Situational Awareness: An Overview*. In W. Rathgeber, K. U. Schrogl, & R. A. Williamson (Eds.), *The Fair and Responsible Use of Space. Studies in Space Policy*. Springer.

Guest, A. N. (2013). Telemetry, Tracking, and Command (TT&C). In J. N. Pelton, S. Madry, & S. Camacho-Lara (Eds.), *Handbook of Satellite Applications*. Springer. https://www.esa.int/Science_Exploration/Space_Science/About_Space_Science (Accessed on 30 January 2023).

ISECG. (2013). *The Global Exploration Roadmap*. International Space Exploration Coordination Group. https://www.globalspaceexploration.org/wordpress/wp-content/uploads/2013/10/GER_2013.pdf (Accessed 5 February 2023).

Joint Chiefs of Staff. (2020). *Joint Publication 3–14: Space Operations*. Joint Chiefs of Staff. https://www.jcs.mil/Portals/36/Documents/Doctrine/pubs/jp3_14Ch1.pdf (Accessed 10 February 2023).

Krasner, S. D. (1999). *Sovereignty: Organized Hypocrisy*. Princeton University Press.

Labrador, V. (2023). *Satellite Communication* .Encyclopaedia Britannica. https://www.britannica.com/technology/satellite-communication (Accessed 10 January 2023)

Logsdon, J. M. (2023). *Space Exploration*. Encyclopaedia Britannica. https://www.britannica.com/science/space-exploration (Accessed on 5 March 2023)

Madry, S. (2013). Introduction and History of Space Remote Sensing. In J. N. Pelton, S. Madry, S. Camacho-Lara (Eds.), *Handbook of Satellite Applications*. Springer.

Moranta, S., Pavesi, G., Perrichon, L., Plattard, S., & Sarret, M. (2018). *Security in Outer Space: Rising Stakes for Europe* (ESPI Report 64). European Space Policy Institute.

NASA. (2014). *Remote Sensing*. National Aeronautics and Space Administration. https://www.nasa.gov/content/remote-sensing/ (Accessed 1 February 2023).

NASA. (2021). *Technology Demonstration Missions Overview*. National Aeronautics and Space Administration. https://www.nasa.gov/mission_pages/tdm/main/overview.html (Accessed 10 February 2023).

OECD. (2011). *The Space Economy at a Glance 2011*. OECD Publishing.

OECD. (2012). *The Space Economy in Figures. How Space Contributes to the Global Economy.* Organisation for Economic Co-operation and Development. OECD Publishing.

Oltrogge, D. L., & Alfano, S. (2019). The technical challenges of better Space Situational Awareness and Space Traffic Management. *Journal of Space Safety Engineering, 6*(2), 72–79.

Payne, G., & Payne, J. (2004). *Key Concepts in Social Research.* SAGE Publications.

Pelton, J. N., & Camacho-Lara, S. (2013). Introduction to Satellite Navigation Systems. In J. N. Pelton, S. Madry, & S. Camacho-Lara (Eds.), *Handbook of Satellite Applications.* Springer.

Pelton, J. N., Madry, S., & Camacho Lara, S. (2013). Satellite Applications Handbook: The Complete Guide to Satellite Communications, Remote Sensing, Navigation, and Meteorology. In J. N. Pelton, S. Madry, S. Camacho-Lara (Eds.), *Handbook of Satellite Applications.* Springer.

Robinson, J. (2011). *Enabling Europe's Key Foreign Policy Objectives via Space* (ESPI Report 30). European Space Policy Institute.

UNCOPUOS. (2018). *Thematic priority 1. Global Partnership in Space Exploration and Innovation.* United Nations Committee on the Peaceful Uses of Outer Space. Series A/AC.105/.

UNDESA. (2022). *World Population Prospects 2022 Summary of Results.* United Nations Department of Economic and Social Affairs. UN DESA/POP/2021/TR/NO. 3. https://www.un.org/development/desa/pd/sites/www.un.org.development.desa.pd/files/wpp2022_summary_of_results.pdf (Accessed 5 January 2023).

UNOOSA. (2018). *European Global Navigation Satellite System and Copernicus: Supporting the Sustainable Development Goals. Building Blocks Towards the 2030 Agenda.* United Nations Office for Outer Space Affairs. https://www.unoosa.org/res/oosadoc/data/documents/2018/stspace/stspace71_0_html/st_space_71E.pdf (Accessed 15 December 2022).

UNOOSA. (2023a). *Benefits of Space: Communication.* United Nations Office for Outer Space Affairs. https://www.unoosa.org/oosa/en/benefits-of-space/communication.html (Accessed 10 November 2022).

UNOOSA. (2023b). *Benefits of Space: Disaster Management.* United Nations Office for Outer Space Affairs. https://www.unoosa.org/oosa/en/benefits-of-space/disasters.html (Accessed 10 November 2022).

UNOOSA. (2023c). *Benefits of Space: Agriculture.* United Nations Office for Outer Space Affairs. https://www.unoosa.org/oosa/en/benefits-of-space/agriculture.html (Accessed 10 November 2022).

UNOOSA. (2023d). "Benefits of Space: Global Health". United Nations Office for Outer Space Affairs. https://www.unoosa.org/oosa/en/benefits-of-space/global-health.html (Accessed 10 November 2022).

UNOOSA. (2023e). *Benefits of Space: Education.* United Nations Office for Outer Space Affairs. https://www.unoosa.org/oosa/en/benefits-of-space/education.html (Accessed 10 November 2022).

US Department of Defense. (2011). *National Security Space Strategy. Unclassified Summary.* http://www.defense.gov/home/features/2011/0111_nsss/docs/NationalSecuritySpaceStrategyUnclassifiedSummary_Jan2011.pdf (Accessed 1 February 2023).

US Department of Transportation. (2023). What is Positioning, Navigation and Timing? https://www.transportation.gov/pnt/what-positioning-navigation-and-timing-pnt (Accessed 10 January 2023)

Weeden, B. (2017). *Space Situational Awareness Fact Sheet.* Secure World Foundation. https://swfound.org/media/205874/swf_ssa_fact_sheet.pdf (Accessed 2 February 2023)

Weeden, B., & Samson, V. (Eds.). (2022). *Global Counterspace Capabilities. An Open Source Assessment.* Secure World Foundation.

Windley, B. F., & Albritton, C. (2021). Earth sciences. *Encyclopaedia Britannica.* https://www.britannica.com/science/Earth-sciences (Accessed 1 March 2023).

Zarkan Cesari, L. (2021). *What's in a Word? Notions of 'Security' and 'Safety' in the Space Context.* United Nations for Disarmament Research (UNIDIR). https://unidir.org/commentary/whats-word-notions-security-and-safety-space-context (Accessed 20 November 2022).

Chapter 4
Comparing Space Actors: An Empirical Assessment

This chapter presents an analysis of the original empirical data collected in light of our theoretical framework. It presents a comparison of major space actors within the two prominent dimensions of capacity and autonomy along with a detailed analysis of the related subdimensions. The actors selected have been among the most active in space in the past decade and have attracted the most attention from general media outlets, practitioners, and scholars alike. They are Australia, Brazil, Canada, China, Europe, India, Israel, Japan, Russia, South Korea, and United States. Building on the results of the measurement exercise, the chapter ultimately provides a comparative assessment of the status and relative position held by the analysed actors within the spacepower matrix.

4.1 Overview

4.1.1 Case Selection and Data

This chapter presents an in-depth assessment of 94 indicators for 11 notable space actors, namely Australia, Brazil, Canada, China, India, Israel, Japan, Russia, South Korea, the United States and Europe, which in this book is treated as a single, distinct player within the international "spacescape" (see Box 4.1).

M. Aliberti et al., *Power, State and Space*, Studies in Space Policy 35,
https://doi.org/10.1007/978-3-031-32871-8_4

Box 4.1 Europe as a single space actor

The decision to treat Europe as a single, unified actor within the international "spacecape" may appear inconsistent with the theoretical framework proposed in this book (i.e., its focus on "state" actors) and clearly raises the major analytical challenge of identifying *Europe's space actorness*. Europe is obviously not a nation-state in the Westphalian sense, nor is it a polity that has emerged from a single, authoritative source like the other space nations (Aliberti et al., 2019). The European Union (EU)—which is the result of a long-standing project of integration initiated more than half a century ago and thus usually regarded as the geo-political entity representing Europe—is itself problematic due to the fact that "it is something more than an intergovernmental organisation, but still less than a fully-fledged European state" (Wallace, 2005). Even more important, when looking at the issue from the perspective of space activities, the EU is neither the only nor the most directly involved actor in the management of European efforts in the field. That role has traditionally belonged to the ESA (Aliberti et al., 2019).

In fact, if a *European space actorness* is to be identified, this can be said to result from the complex interplay of three distinct constituencies, specifically ESA, the EU and the individual member states of the two organisations, which despite having a common basis, do not coincide. Each member state has a governance structure and participates in several organisations engaged in space matters. Among 27 EU member states and 22 ESA members, 19 belong to both organisations.

The interplay between national, intergovernmental and communitarian frameworks has created institutional misalignment, making decision-making processes inherently cumbersome.

Notwithstanding the obvious limitations, this study finds it practical to empirically analyse Europe as a single, though *sui generis*, space actor. Manifold the rationales informing this decision.

For one thing, the decision stems from the fact that current space activities and capabilities in the European context, even those led by individual countries, are so closely interlinked with each other, that it would be barely impossible to create clear-cut boundaries between individual capacities, assets or programmes and extricate most of the indicators we consider here. Analysing individual European states within the matrices we propose would result in the misleading placement of individual actors that, in reality, work closely together in the space domain.

Second, the European space industrial base is highly consolidated, with a high degree of vertical integration across the continent. Because of the fundamental characteristics of the space business, there has been a natural tendency towards the emergence of vertically integrated space enterprises (Hayward, 2011). European space companies went through a consolidative transformation through mergers and acquisitions in the 1990s, which led to two space company groups at the prime contractor level, Airbus Defence & Space and Thales Alenia Space. These two groups are complemented by OHB as another major system integrator (Al-Ekabi, 2016). As a result, the European space sector is nowadays more advanced than any other industrial sector in terms of supply chain integration.

Third, over the past 20 years the path towards a more integrated European space governance has steadily deepened. Besides the ongoing enlargement of ESA towards all the EU Member States not yet members of the Agency, this is demonstrated by the closer cooperation between ESA and the EU. The two organisations have recognised that both parties have specific complementary and mutually reinforcing strengths and need each other to fulfil public policy objectives and provide a more coherent framework for space activities in Europe (Aliberti, 2015). Whereas the EU holds legislative power and political clout, ESA is equipped with technical and operational expertise. Since the publication of the first EC/ESA Framework Agreement in 2004, the EU and ESA have thus committed themselves to working together for the implementation of mutually beneficial space projects, avoiding duplication of efforts and optimizing resource allocation. Currently, the EU delegates a significant part of its space budget to ESA for pan-European programmes like Copernicus and Galileo.

Fourth, notwithstanding the institutional mismatch between the various actors and the core interests laid down in their respective strategies, important convergences and a set of common objectives that are shared across all stakeholders composing the European space governance have increasingly consolidated in recent years. This consolidation was eventually reflected in the Joint EU/ESA Statement on "Shared Vision and Goals for the Future of Europe in Space" adopted on 26 October 2016. Despite not being legally-binding, this document is key, as it not simply represents an agreement between ESA and the EU on a number of common goals for the European space sector, but—having received explicit support also from member states—represents a European-wide convergence on a shared strategic vision for space activities (Aliberti et al., 2020). This joint statement was followed by the signature of the *Financial Framework Partnership Agreement* (FFPA) between the EC and ESA, signed on 22 June 2021. The document defines the roles, responsibilities and obligations of the EC, the EUSPA, and ESA with regard to each component of the space programme, deals with the shared costs for European space programmes, and provides the coordination and control mechanisms.

In light of all these considerations, it seems appropriate to assess Europe's position within the spacepower matrix as that of a unified, though *sui generis*, internationally-acting body.

The reasons for circumscribing the analysis to these specific players are multifaceted. First, during the period under review (which spans from 2011 to 2020), these actors have been the most active worldwide in the fields of space technology, exploration, and utilisation. Table 4.1 provides an overview of the space activities of the 11 countries included in this analysis in comparison to the rest of the world. This overview encompasses data on several objective indicators including the number of active satellites, the number of satellites launched in the decade, the number of launches, launch cadence, and total mass launched or number of space exploration missions. Clearly, our 11 cases represent the vast majority of all space activities carried out globally between 2011 and 2020.

Second, and closely related, interest for these countries' space activities has been steadily growing during the last ten years: space feats in nations as diverse as Australia, China, India, and Brazil have attracted attention from unspecialised media outlets, practitioners, and the academic community alike, with dozens of publications coming to the fore and providing invaluable sources of information for conducting comparative analysis.

Third, the countries selected for the examination are representative of the "Global West"—which comprises the United States and its allies in Europe, Israel in the Middle East, and Australia, Japan, and South Korea and the Pacific region—and the most powerful emerging economies of the BRICS, namely Russia, China, India, and Brazil.

Fourth, there have been objective difficulties in expanding the comparative base by including such new entrants as Argentina, Iran, Nigeria, South Africa, Turkey the United Arab Emirates due to the limited availability of both hard data on their

Table 4.1 Space activities overview in the period 2011–2020

Country of operator/owner	No. of satellites launched	No. of launches	Launch cadence	Total mass launched (tons)	No. of space exploration missions
Australia (AUS)	15	0	0	16.2	0
Brazil (BRA)	10	0	0	21	0
Canada (CAN)	41	0	0	39.5	4
China (CHN)	457	240	24	584.1	19
Europe (EUR)	459	96	9.6	272.6	16
India (IND)	65	44	4.4	92.2	3
Israel (ISR)	17	3	0.3	19.9	0
Japan (JAP)	118	39	3.9	240.2	15
South Korea (KOR)	25	1	0.1	22.7	3
Russia (RUS)	259	236	23.6	832.3	83
United States (US)	2317	240	24	1396.6	70
Rest of the World	201	28	2.8	215.7	2
Total	3984	927	92.7	3953	215

national space activities and country experts (both inside and outside the polity) willing undertake the survey. The exclusion of these new entrants—which would be expected to fall within the "emerging space nations" quadrant—naturally skews the distribution towards higher values but lays the ground for a more focused comparison among more established space actors. Future comparative analyses, as well as future country-specific works on spacepower should look at the space activities of these "emerging space nations" in more detail.

As presented in Chap. 3, the empirical assessment devised in this study draws on both quantitative and qualitative data: quantitative data have been retrieved from various ad hoc sources, while qualitative data result from answers to a survey administered to prominent space specialists. Our quantitative data have been processed, standardised, and integrated to ensure full comparability. Similarly, when submitting answers to our survey, country-experts were expected to provide evidence-based judgments and/or to refer to objective indicators.

4.1.2 *Positioning of Space Actors in the Spacepower Matrix: Outline*

The ultimate goal of this chapter is to use a multitude of empirical data collected in order to generate a spacepower matrix that can be utilised to effectively compare space actors in terms of their spacepower. In order to generate this matrix, and following our theoretical framework (see Sect. 2.3 in Chap. 2), we generate two sub-matrices for each of the two dimensions that we consider fundamental to measure spacepower: capacity and autonomy. In the next section we discuss in-depth the capacity dimension and the capacity sub-matrix. We then move on to a detailed discussion of the autonomy dimension and the autonomy sub-matrix. Once both sub-matrices are established, we are able to create the space power matrix, place each country included in the analysis within the matrix, and briefly discuss country results individually.

We ultimately find that space actors that enjoy the status of space power as conventionally understood also possess higher-than-average levels of both capacity and autonomy as we define them. Figure 4.1 shows the final results of our comparative analysis.

The United States, China and Russia emerge as the three indisputable leading space powers. They are the only three countries to score in the top quartile of both dimensions. Europe follows closely, with Japan and India at a distance. Interestingly, none of the countries in our sample fits the emerging space nations quadrant (low capacity and low autonomy) or the quadrant named skilled spacefaring nations (highly capable space nations with rather limited autonomy). On the contrary, with the exception of Europe, most states appear to be more autonomous (especially in the hard autonomy realm, as we will discuss later) than capable—thus falling in the autonomous spacefaring nations quadrant.

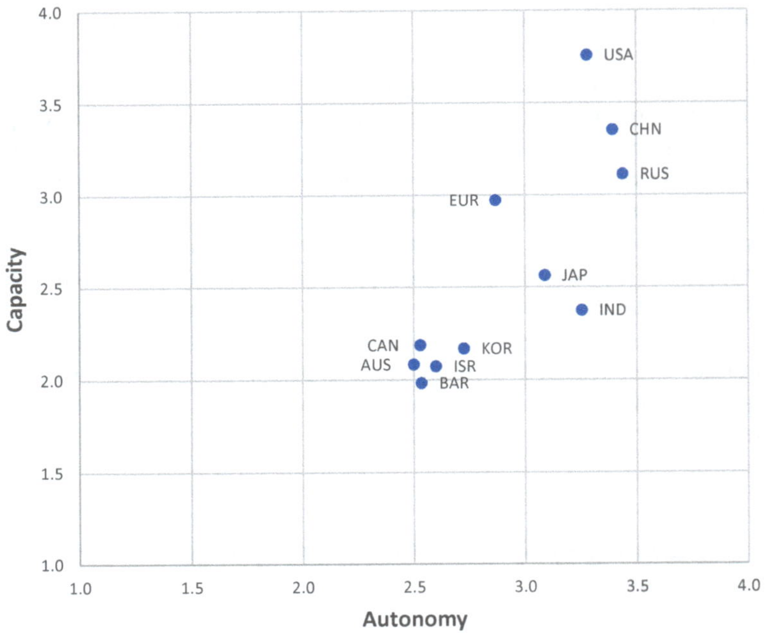

Fig. 4.1 Positioning of space actors in the spacepower matrix

4.2 Measuring and Comparing Capacity

In our taxonomy, "capacity" denotes a state's ability to *implement* a space strategy and to achieve the related goals in the political, diplomatic, military, economic, or social realms. We break down the concept of capacity into two sub-dimensions, "hard capacity" and "soft capacity." Hard capacity encompasses the full spectrum of material assets and technological skills which enable an actor to operate in, through, and from space; it has been measured by hard data and objective indicators. Soft capacity, in turn, denotes the ability of an actor to effectively integrate its assets and skills in a variety of areas, from domestic socio-economic infrastructures to foreign and security policies. Its measurement is more complex and nuanced, and we turned to the subjective assessment of country experts to develop it.

4.2.1 Measuring Hard Capacity

What we call hard capacity stretches across four prominent macro-areas: "satellite applications," "science and exploration," "space safety and security," and "enabling and support." Each macro-area in turn covers four areas, for a total of 38 indicators. Values for indicators have been retrieved from the database of the European

Space Policy Institute (ESPI) and other open-source datasets including the Center for Strategic and International Studies (CSIS, 2019), the Science and Technology Policy Institute—Institute of Defense Analyses (IDA, 2020), and the Union of Concerned Scientists (UCS, 2021). Scores run on a scale ranging from 1 (lowest) to 4 (highest). Figure 4.2 summarises the results of our hard capacity analysis for all 11 space actors included in our study.

As it might be expected, the United Stated, China and Russia display the highest scores in all macro-areas, closely followed by Europe. The greatest space power of all (the US) leads everywhere and also sports the most balanced internal composition among the macro-areas: in particular, in the US the macro-area of Science & Exploration receives virtually the same scores as the macro-areas of Satellites Applications and of Enabling & Support, with Space Safety & Security following at a distance. While space safety and security lags behind the other macro-areas considered in all countries except Canada, Australia, and Russia only the United States attributes to science and exploration the same relevance as satellites applications, which is the single most developed macro-area everywhere else. This attention given to science and exploration is probably one of the reasons why the US is a clear world leader

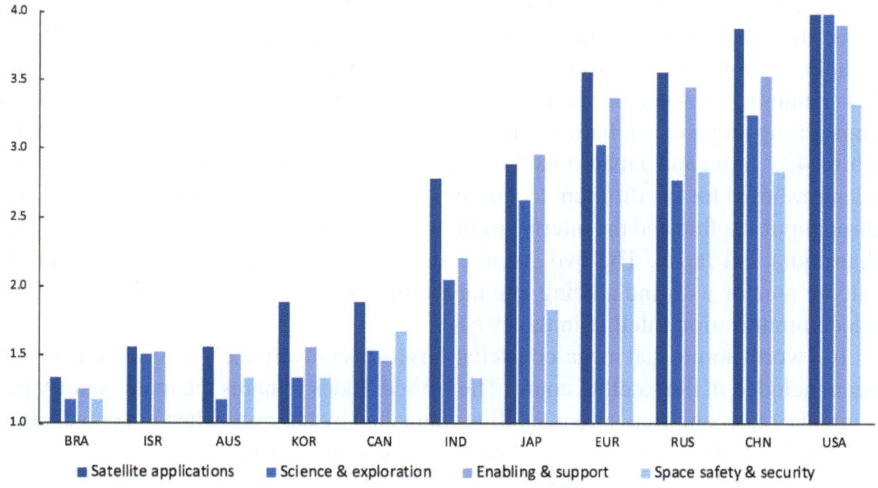

Reference data

	BRA	ISR	AUS	KOR	CAN	IND	JAP	EUR	RUS	CHN	USA
Satellite applications	1.3	1.6	1.5	1.9	1.9	2.8	2.9	3.6	3.6	3.9	4.0
Science & exploration	1.2	1.5	1.2	1.3	1.5	2.1	2.6	2.8	3.0	3.2	4.0
Enabling & support	1.3	1.5	1.5	1.6	1.5	2.2	3.0	3.4	3.5	3.6	3.9
Space safety & security	1.2	1.0	1.4	1.4	1.7	1.3	1.9	2.2	2.9	2.9	3.4
Hard Capacity Index *(avrg)*	**1.2**	**1.4**	**1.4**	**1.6**	**1.7**	**2.1**	**2.5**	**3.0**	**3.3**	**3.4**	**3.8**

Fig. 4.2 Inside hard capacity

in the field. In the next pages, we present a more detailed look at each of the four macro-areas considered.

4.2.1.1 Hard Capacity: Satellite Applications

Overall, in the macro-area "satellite applications," the United States (4.0) stands out as the world leader followed by China (3.89): both countries' endeavours in the decade from 2011 to 2020 have showcased how space assets have become an integral part of national economy and infrastructures with significant technical capacity developed and achieved in terms of satellites launched and operated. Russia (3.56) and Europe (3.56) follow, the former rewarded by its military use of space, especially in remote sensing, and the latter by its commercial telecommunication satellites operated by privately owned companies like the SES Group, Eutelsat and OneWeb. Japan (2.89) and India (2.78) result fair solid actors in the satellite applications macro-area in general, followed then by South Korea (1.89) and Canada (1.89).

Looking more in detail at the results, the macro-area "satellite applications" was divided into three areas: telecommunications, remote sensing, and navigation, each of which was measured on the basis of three indicators, as shown in Fig. 4.3.

In the "telecommunication satellites" area the United States (4.0) leads the sample, with China and Europe falling just behind it (at 3.7), penalised by the lower number of satellites launched over the period from 2011 to 2020 (without considering mega-constellations, the US launched 217 satcom satellites, Europe 92, China 42). India and Japan obtain the same score as Australia (2.3). Australia has been rewarded for the different technologies it has deployed via satellite (e.g., high-throughput satellite and the internet of things), despite having launched less satellites than India and Japan. The two Asian countries have a longstanding history in the satcom area of with India being among the first Asian countries to have a national telecommunication satellite in the 1970s.

It is worth noting that mega-constellations (especially SpaceX's Starlink) have not been included in the satellite count. The United States remains the main country per number of telecommunications satellites launched by far, even when not considering the 1600+ Starlink satellites launched by SpaceX since 2019.

Looking at the "remote sensing" area, China and the United States (4.0) emerge as the top scorers, immediately followed by Russia (3.7). The three countries (albeit with relevant differences in terms of number of satellites launched in the decade, with the US having launched 592, China 191, and Russia 36), have committed significantly to this segment of the sector, with a vast array of satellites and a relevant number of institutional users which power the demand for satellite imagery. Europe is following the trio (3.3), slightly penalised for the lack of military satellites to provide early warning and SIGINT support from space during the considered timeframe. Japan (3.0), India (3.0), and South Korea (2.7) also score relatively well, thanks to the deployment of a vast array of sensors. Australia (1.0) has not launched a single remote sensing satellite in the decade 2011–2020, thus receiving the lowest scores.

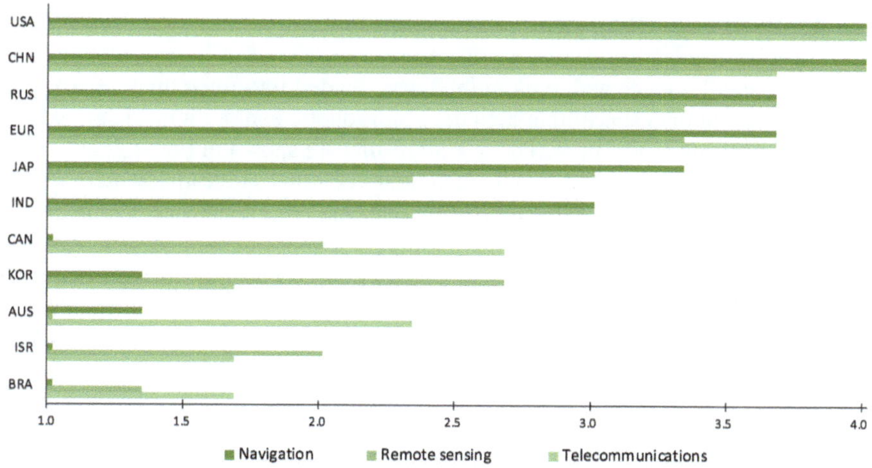

Reference data

	BRA	ISR	AUS	KOR	CAN	IND	JAP	EUR	RUS	CHN	USA
Telecommunications Number of satellites, Mission diversity, Performance (technological assortment)	1.7	1.7	2.3	1.7	2.7	2.3	2.3	3.7	3.3	3.7	4.0
Remote sensing Number of satellites, Mission diversity, Performance (variety of sensor types)	1.3	2.0	1.0	2.7	2.0	3.0	3.0	3.3	3.7	4.0	4.0
Navigation Coverage, Augmentation (space-based), Performance (accuracy)	1.0	1.0	1.3	1.3	1.0	3.0	3.3	3.7	3.7	4.0	4.0
Satellite Application *(avrg)*	**1.3**	**1.6**	**1.5**	**1.9**	**1.9**	**2.8**	**2.9**	**3.6**	**3.6**	**3.9**	**4.0**

Fig. 4.3 Satellite applications

Finally, for what regards "positioning, navigation and timing," the United States and China score again the highest (both 4.0). The two powers have deployed full GNSS constellations, a space-based augmentation system, and have been able to achieve significant (sub-metre) accuracy for their systems. Europe and Russia (3.7 both) follow, with the latter penalised by to the relatively lower accuracy of its GLONASS system, and the former by the fact that the "Galileo" constellation was still fully deployed at the end of the considered timeframe (2011–2020). India (3.0) and Japan's (3.3) scores are respectable due to their deployment of both the RNSS NaVIC and its SBAS component. Not surprisingly, the other countries considered score lower given the lack of national navigation satellite systems. Australia's intention to develop a SBAS, in collaboration with New Zealand (LINZ, 2022), granted the country a slightly higher score (1.3) that other nations with limited capacity in this area.

4.2.1.2 Hard Capacity: Science and Exploration

The macro-area "science and exploration" was divided into three areas: space and Earth sciences, human spaceflight activities, and robotic exploration, each of which is made of dedicated indicators. The overall results are shown in Fig. 4.4.

In the macro-area as a whole, the United States continues to lead the international space pecking order by a significant margin over the other countries. Compared to the other countries the United States has shown consistency over the decade for what concerns the launch of "astronomy and space science" missions as well as "earth science" missions. The Americans have in fact launched 67 satellites over a decade for this purpose, more than the total combined number of satellites launched for this purpose by the Europeans (31), the Chinese (16) and the Russians (14).

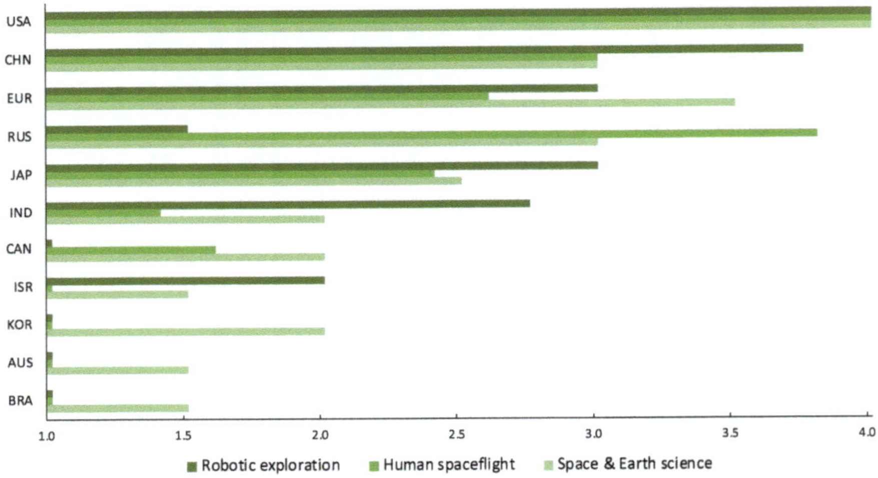

Reference data

	BRA	AUS	KOR	ISR	CAN	IND	JAP	RUS	EUR	CHN	USA
Space and Earth science No. of Space science missions, No. of Earth Science	1.5	1.5	2.0	1.5	2.0	2.0	2.5	3.0	3.5	3.0	4.0
Human spaceflight Astronauts, Extra-vehicular activities (EVAs), Cargo vehicles/programs, Crewed vehicle/programs, Space stations infrastructures	1.0	1.0	1.0	1.0	1.6	1.4	2.4	3.8	2.6	3.0	4.0
Robotic exploration No. of missions launched in 2011-2020, No. of operational missions, Diversity of destination, Mission architecture variety	1.0	1.0	1.0	2.0	1.0	2.8	3.0	1.5	3.0	3.7	4.0
Science & Exploration *(avrg)*	**1.2**	**1.2**	**1.3**	**1.5**	**1.5**	**2.1**	**2.6**	**2.8**	**3.0**	**3.2**	**4.0**

Fig. 4.4 Science and exploration

It is nevertheless worth acknowledging that this type of mission is international in scope, with data shared across borders, and scientists from around the world working together with the same datasets. It makes, therefore, sense for smaller countries to base their own science missions on foreign, more advanced spacecraft, maybe contributing with payloads, sensors, or specific expertise to the various missions.

As for "human spaceflight," the United States (4.0) and Russia (3.8) unsurprisingly lead the group of considered countries. The two countries represent the current backbone of global crewed spaceflight, thanks to their experience, orbiting infrastructure, and crewed launch vehicles. China's (3.0) score is lower just due to the country's relatively novel engagement with human spaceflight and lack, for instance, of EVAs activities during the considered timeframe—China's first ever EVA happened in 2008, with the second just in 2021 (Jones, 2021). Undoubtedly, current Chinese feats such as the assembly of the new Tiangong Spaces Station will positively impact such indicator in the short-to-medium term. Europe (2.6) follows together with Japan (2.4), with the Europeans being rewarded by their contribution to crewed capsules programmes as Orion (ESA, 2021a) and both countries standing over Canada (1.6), Australia (1.0), Brazil (1.0) and South Korea (1.0) thanks to their contribution to the infrastructure and programme of the International Space Station (ISS). Among the remaining countries, Canada stands out thanks to notable contributions including the Canadarm-2, as well as to the sensible presence of Canadian astronauts aboard the station.

Finally, "robotic exploration" showcases the United States' (4.0) prowess, as well as China's (3.7) rise as a relevant space actor in the area. Robotic exploration is a costly, technologically demanding, and highly risky endeavour; this explains why in the last decade only the United States has managed to launch almost one new mission per year across the solar system, targeting a variety of destinations including Mars, asteroids, and dwarf planets (Ceres and Pluto), with a wide variety of spacecraft architectures, including a sample return mission and the technology demonstrator "Ingenuity." China follows the United States with more than a new mission launched every other year, with the various Chang'e missions leading up to the successful Chang'e 4 and Chang'e 5 (The Planetary Society, 2021a, b), as well as the launch of the ambitious Tianwen-1—mission that included orbiter, lander, and the rover Zhurong, just recently successfully deployed on the surface of Mars (The Planetary Society, 2021d). Europe (3.0), Japan (3.0), and India (2.8) have launched less missions in the period 2011–2020, albeit what has been launched is surely ambitious on their side Europe has been penalised by the delay of the Rosalind Franklin rover, whose launch has been postponed from 2020 to 2022 (O'Callaghan, 2020). Surprisingly, Russia's (1.5) robotic exploration efforts have been almost null in the last decade, with the only independent attempt being made with the launch of the unsuccessful Fobos-Grunt (NASA, 2021a) and the contributions to the ESA-led ExoMars missions.

Overall, as for "science and exploration," the United States (4.0) and China (3.25) result in the two most prominent actors, with Russia (2.77) being outclassed by China's ambitious feats in robotic exploration, and the Russian fatigued human spaceflight programme being the only indicator where the country still scores above the Chinese. Europe (3.03) and Japan (2.63) also fare well, thanks to, both countries' participation in the ISS programme, as well as ambitious missions conducted during the period of reference such as the visit of the Comet 67P/Churyumov-Gerasimenko by ESA's Rosetta mission (ESA, 2021b) or to the Asteroid Ryugu by Hayabusa-2 (JAXA, 2021).

4.2.1.3 Hard Capacity: Space Safety and Security

The "space safety and security" macro-area, embraces two areas: "Space Situational Awareness" (SSA) and "Counter-space." In the macro-area as a whole the United States (3.34) leads, followed by China and Russia, both scoring 2.84. Europe (2.17) is the only other actor that scores higher than 2. The overall prevalence of the US in this macro-area comes as no surprise as the country is particularly rewarded by its efforts in SSA. Russia and China (both scoring 2.84) follow, being instead prized more by their scores in counter-space. Europe owes its thirst post to its score in SSA, and space weather particularly. Japan (1.84) and Canada (1.67), scoring respectively almost half of the United States' score, are worth mentioning thanks to a displayed well-balanced mix of SSA capabilities (Fig. 4.5).

As for the "space situational awareness" area—which is articulated in the three entries "space surveillance and tracking" (SST), "near-earth objects" (NEOs) and "space weather"—the United States (4.0) stands out again as the country with the highest capacity. The United States has a well-developed global network of observatories, thanks to its military presence and bilateral relations such as that with the five-eyes countries. The United States has also the best funded programmes for what concerns space weather and NEOs tracking.

Europe (3.0) follows the US, with a lower score in SST but demonstrating good capacity for what concerns NEOs and space weather monitoring. Europe's score will likely improve, given the region's efforts in joint European Space Traffic Management and SST solutions (European Commission, 2022). China and Russia (each scoring 2.7) follow, both countries having developed advanced capabilities in SST (both radar and optical), but lagging behind for what concerns NEOs. The other countries score lower. Capacity in SSA is space-specific, and it can be considered essential for a country once it owns many assets in space, with India (1.3) as the sole exception among major space actors, being very active but not having SSA capabilities. Canada (2.3) and Australia (1.7) have some SST capabilities which significantly contribute to their partnership with the United States, both countries having also developed experimental in-orbit SSA capabilities.

Evaluating the "counter-space" area is somewhat puzzling, given its intrinsic opaqueness and the difficulty for governmental and private actors alike to assess the nature of orbital operations—such as Rendezvous and Proximity Operations (RPO) manoeuvres—or to gain specific information on foreign arsenals—as the number of electronic and cyberwarfare equipment, direct ascent or co-orbital ASAT weapons. China (3.0) and Russia (3.0) score the highest, with China being the country that has likely deployed the most effective capabilities, based on publicly obtainable

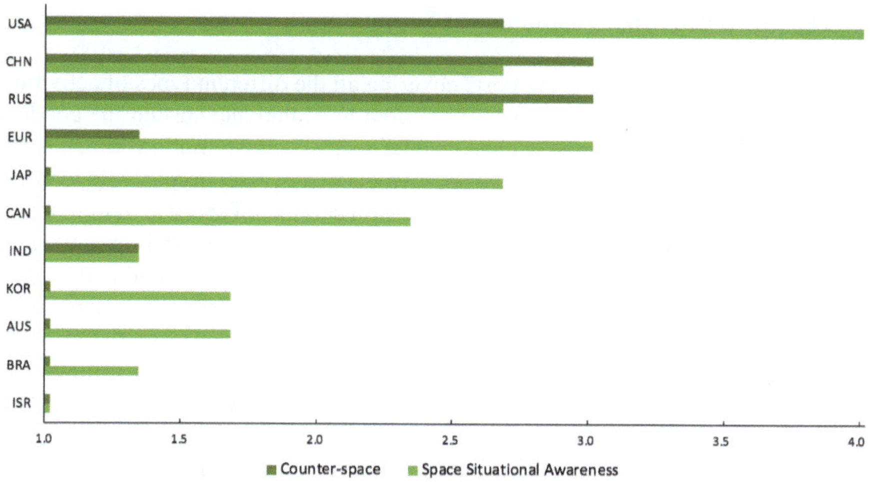

Reference data

	ISR	BRA	AUS	KOR	IND	CAN	JAP	EU	RUS	CHN	USA
Space Situational Awareness Space surveillance and tracking, Near Earth objects, Space weather	1.0	1.3	1.7	1.7	1.3	2.3	2.7	3.0	2.7	2.7	4.0
Counter-space R&D of counter-space capabilities, Deployment of counter-space capabilities, Testing of counter-space capabilities, Deployment of counter-space capabilities	1.0	1.0	1.0	1.0	1.3	1.0	1.0	1.3	3.0	3.0	2.7
Space Safety & Security *(avrg)*	**1.0**	**1.2**	**1.4**	**1.4**	**1.3**	**1.7**	**1.9**	**2.2**	**2.9**	**2.9**	**3.4**

Fig. 4.5 Space safety and security

intelligence and observation of orbital manoeuvres. Space assets of both countries have for instance been observed conducting unannounced RPOs—e.g., China's SJ class of missions or Russia's Kosmos 2504, Kosmos 2542, and Luch/Olymp-k satellites (Martin et al., 2021). This time the United States (2.7) follows, slightly penalised for the lack of information on its deployed and tested capabilities, with the government not disclosing information and sources less available from open-source intelligence. All the other countries score the minimum, with India (1.3) distinguishing itself for its research and testing on direct-ascent and direct-energy anti-satellite weapon—a clear example being the ASAT test conducted in March 2019 (MEA, 2019)—and Europe (1.3)—especially thanks to France—working to experiment and deploy direct energy weapon and means to conduct electronic warfare.

4.2.1.4 Hard Capacity: Enabling and Support

Our macro-area "enabling and support" covering all the different types of activities, technical capabilities and infrastructure required to enable and sustain the conduct of any space mission. The most evident enabler is space transportation, which is the gateway for any orbital activity and the deployment of any asset in space. Beyond the launch system itself, also critical for the successful conduct of space activities are all the infrastructural elements that support the launch, deployment and operations of space assets. These include the launch facilities, the mission command and control centres, the different land-, sea- and space-based stations, tracking stations and, more broadly, all facilities performing (TT&C) functions.

Our analysis overall shows the United States (3.92) leading, closely followed by China (3.54) and Russia (3.46), with Europe (3.38) just slightly after the latter two (see Fig. 4.6).

In the area of "space transportation," the United States scores the maximum, thanks to the globally acknowledged variety, performance, and reliability of its launchers, as well as thanks to the incredible launch cadence of the last decade, particularly influenced by SpaceX's efforts (101 launches of the 240 launches conducted by the US in 2011–2020). China (3.5), Europe (3.5), and Russia (3.5) follow with the same score, with Europe being rewarded by its launchers' reliability (96 launches with 2 failures and 2 partial failures) and Russia and China by their higher launch cadence (240 and 236 attempted launches respectively). Japan (3.0) is rewarded by both its launchers' reliability and performance of the now retired H-IIB, while India's (2.5) scores are positively impacted by its flawless launch success rate (44 successful launches out of 44 launches in total). Among the other considered countries, only Israel (1.2) and South Korea (1.2) had active launchers in the period 2011–2020, but their launch rate (in 2011–2020, three launches for Israel, and one for South Korea) was too low to provide meaningful data to evaluate the "launch cadence" and "launcher's reliability" indicators.

As for "ground operations," the United States (4.0) and China (4.0) lead with top scores with respect to both the TT&C infrastructure and launch facilities, followed by Russia (3.5), and then Europe (3.0) and Japan (3.0)—the former being rewarded by its ESTRACK deep space network (ESA, 2021b), while penalised by its low number of active launch facilities (having launched from just one single spaceport, the Guyana Space Centre (GSC), in the decade considered), which could become a bottleneck if Europe ever decided to increase its space activities. India (2.5), South Korea (2.0), and Israel (1.5) are similarly penalised by their low number of launch facilities, although this impacts less these countries, compared to Europe, given their number of space activities. Australia (1.0), Brazil (1.0), and Canada (1.0) are all penalised by the lack of extensive, globally spanning sovereign TT&C network (not owning tracking ships nor national Satellite Data Relay systems), as well as by the lack of active orbital launch facilities. The three countries have pledged an increase in activities, participation in the Artemis Accords, and the development of indigenous launch capabilities, so they will likely increase these scores in the coming decade.

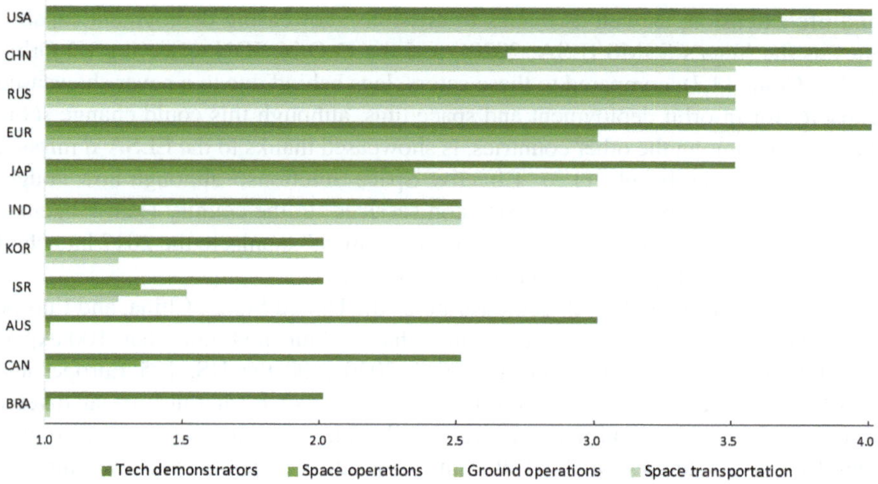

Reference data

	BRA	CAN	AUS	ISR	KOR	IND	JAP	EUR	RUS	CHN	USA
Space transportation	1.0	1.0	1.0	1.2	1.2	2.5	3.0	3.5	3.5	3.5	4.0
Launch cadence, Launchers' variety, Performance, and Reliability											
Ground operations	1.0	1.0	1.0	1.5	2.0	2.5	3.0	3.0	3.5	4.0	4.0
TT&C infrastructure, Launch facilities											
Space operations	1.0	1.3	1.0	1.3	1.0	1.3	2.3	3.0	3.3	2.7	3.7
Active-debris removal (ADR) capabilities, Life extension services, In-space deployment capabilities											
Tech demonstrators	2.0	2.5	3.0	2.0	2.0	2.5	3.5	4.0	3.5	4.0	4.0
No. of tech demonstrator missions, Nature of tech demonstrators											
Enabling & support *(avrg)*	**1.3**	**1.5**	**1.5**	**1.5**	**1.6**	**2.2**	**3.0**	**3.4**	**3.5**	**3.6**	**3.9**

Fig. 4.6 Enabling and support

In the "space operations" area, the United States (3.7) leads thanks to civil, military, and commercial missions that help the country score in the indicators on "life extension" (e.g., SpaceLogistics'/Northrop Grumman's MEV missions), "deployment" (e.g., Nanorack's Bishop, ISS), "space tugs" (e.g., Sherpa, Vigoride), and "space planes" (X-37B) technologies, as well as the development of the robotic servicing mission "OSAM-1" (NASA, 2021b). Russia (3.3) follows suit thanks to its demonstration of in-orbit deployment with the military satellite Kosmos-2519 (Krebs, 2021), as well as the capacity to deploy satellites by means of EVAs from the ISS, as well as feats as the July 14, 2017, Soyuz launch that orbited and deployed 70+ satellites (Clark, 2017). Europe (3.0) follows thanks to a well-balanced set of in-orbit validations for technologies that could be used for active debris removal and life extension, as well as thanks to missions showcasing deployment capabilities as the Vega's SSMS (ESA, 2020), and the D-Orbit ION space tug (D-Orbit, 2021). China (2.7), compared to these actors, lags behind due to a never showcased capacity for in-orbit deployment and space tugs, although this could change soon. Japan (2.3), among the other countries, is showcased thanks to the ELSA-d mission (Atroscale, 2021). Israel's (1.3) "Effective Space Solutions," although now bought out, attributed one point in "life extension services" to the country (Van Wagenen, 2015) while India (1.3) was rewarded in "deployment" thanks to the 2017 launch of the PSLV with 104 spacecrafts on board (ISRO, 2021).

Finally, in the area of "tech demonstrators" the United States, China, and Europe score the highest (4.0)—the three countries having launched more than 100 experimental satellites each in the decade 2011–2020 (300 the US, 125 Europe, and 114 China). In addition, the organisations operating such satellites are as diverse as to include academia, military, governmental institutions, and commercial companies. Russia (3.5) and Japan (3.5) follow the three actors, penalised by a lower number of experimental satellites launched (20 for Russia, 50 for Japan).

India (2.5) may surprise for the low number of experimental spacecraft launched compared to other prominent space actors, with only 6 tech demonstrators launched in the decade 2011–2020. Australia (3.0), finally, scores high for the involvement of many different entities from government, military, commercial, and academia, although the number of experimental spacecraft launched remains overall low (5 over a decade).

4.2.2 Measuring Soft Capacity

Soft capacity denotes the ability to effectively utilise and integrate assets and expertise in national policies, infrastructure, and activities. It stretches across two major macro-areas: "foreign and security policies" and "socio-economic policies" which in turn encompass three and four areas respectively (security, defence, and foreign policy belong to the first macro-area, while environment and resources, infrastructures, development and growth, and civil society belong to the second).

Overall, the Soft Capacity Index covers a total of 26 entries, each of which has been assigned scores based on answers to our experts' survey and running on a scale ranging from 1 (lowest) to 4 (highest).

Importantly, as explained in Chap. 3, scores attributed to each entry have been averaged to form a single value for each country. Similarly, scores attributed to each area also form a single averaged value. The same applies to scores attributed to each macro-area.

The overview in Figure 4.7 shows that the United States, China, and Russia score the highest values (>3) in the global soft capacity index (average), followed closely by Europe.

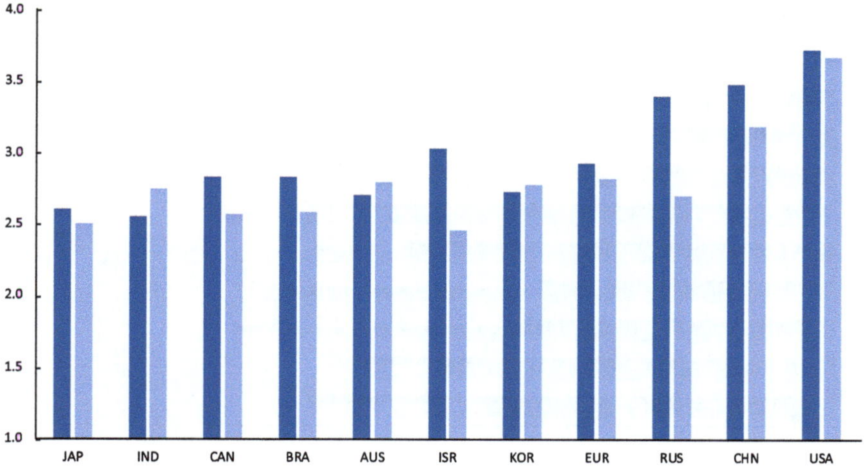

Reference data

	JAP	IND	CAN	BRA	AUS	ISR	KOR	EUR	RUS	CHN	USA
Foreign & security policies	2.6	2.6	2.8	2.8	2.7	3.1	2.8	2.9	3.4	3.5	3.7
Social & economic policies	2.5	2.8	2.6	2.6	2.8	2.5	2.8	2.8	2.7	3.2	3.7
Soft capacity index *(avrg)*	**2.6**	**2.7**	**2.7**	**2.7**	**2.8**	**2.8**	**2.8**	**2.9**	**3.1**	**3.4**	**3.7**

Fig. 4.7 Inside soft capacity

In all countries (except India, Australia and South Korea) the "Foreign & Security" macro-area scores higher than the "Social & Economic" macro-area. This is particularly evident in the case of China, Russia and Israel, (reflecting the overwhelming importance of the military use of space for these countries) while the two macro-areas appear tendentially well balanced in the US, Europe, South Korea and Australia which testify of a more complex and differentiated use of space technologies in these counties (Fig. 4.7).

In the next pages, we present a more detailed look at each of the two macro-areas considered.

4.2.2.1 Soft Capacity: Foreign and Security Policies

The use of space for "foreign and security policies" is more tied to the tradition of space activities, being the sector closely connected to strategic and military considerations since the first space endeavours and the establishment of an industrial apparatus that could sustain them. In this macro-area, respondents have assigned the highest scores to the United States (3.71), closely followed by China (3.51) and Russia (3.4) (Fig. 4.8).

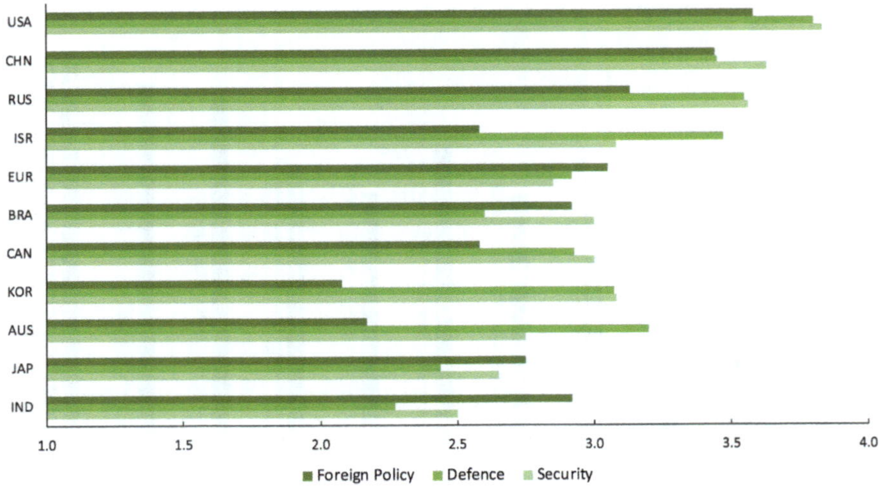

Reference data

	IND	JAP	AUS	KOR	CAN	BRA	EUR	ISR	RUS	CHN	USA
Security National security policies, Surveillance and monitoring, Disaster prevention and management	2.5	2.7	2.7	3.1	3.0	3.0	2.8	3.1	3.6	3.6	3.8
Defence National military strategy, Prevention and deterrence, Military operations	2.3	2.4	3.2	3.1	2.9	2.6	2.9	3.5	3.5	3.5	3.8
Foreign Policy Space for diplomacy, Space diplomacy, Foreign aid and international initiatives, International prestige	2.9	2.8	2.2	2.1	2.6	2.9	3.0	2.6	3.1	3.4	3.6
Foreign & security policies *(avrg)*	**2.6**	**2.6**	**2.7**	**2.8**	**2.8**	**2.8**	**2.9**	**3.1**	**3.4**	**3.5**	**3.7**

Fig. 4.8 Foreign and security policy

Responses for the use of space for and in the "security" area average the highest among the three areas of this macro-area (3.15), a consequence of many countries using space often or very often when it comes to security, according to the respondents of the survey. The survey acknowledges the United States (3.8), China (3.63), and Russia (3.6) as the countries that most often deploy space capabilities for security applications, followed by Korea (3.08) and Israel (3.08)—two countries allied to the United States and tightly connected to the US projection of space forces globally. Brazil and Canada also score relatively high (both 3.00), while, significantly enough, the space actors which score the lowest—Europe (2.8), Australia (2.7), Japan (2.7), and India (2.50)—are all penalised by the lack of inclusion of space in national security policies, something that is slowly changing at the European level, as well as in Japan (Aliberti & Hadley, 2020).

The United States (3.8), Russia (3.5), and China (3.5) lead the other countries in the "defence" area, followed by Israel (3.5), Australia (3.2), and Korea (3.1)—with the respondents acknowledging the importance that space plays both for national military strategies and the conduction of military operations, two indicators where the six countries particularly distinguish themselves as being the ones scoring higher than "3.0." These countries have a significant portfolio of openly military or dual space assets, with the US having launched 161 of such satellites in the period 2011–2020, China 145, and Russia 91, while Israel and Australia heavily relying on partnerships and data from commercial operators and allies as the United States.

Finally, as for the use of space in "foreign policy," the countries scoring the highest are the United States (3.6), China (3.4), and Russia (3.1). Responses suggest that Brazil and India follow (both 2.92), the former rewarded by sensible activity in international space fora and the use of space when dealing with foreign aid and international initiatives, the latter particularly benefiting from the use of space for boosting international prestige. The lowest scoring countries are Israel (2.6), Australia (2.2), and South Korea (2.1)—three actors that only recently have decided to (re)boost their image as spacefaring country on the international arena, and that might become more active in international fora such as UN COPUOS, or to use space to boost international prestige. In this direction, the Israeli ill-fated mission Beresheet provided an opportunity for this, and a second mission is now being planned (The Planetary Society, 2021c); while Australia's participation in the Artemis Accords and lunar missions will likely play a similar role, and the same can be said for South Korea's new lunar plans (Park, 2021).

4.2.2.2 Soft Capacity: Socio-Economic Policies

The macro-area named "socio-economic policies" is characterised by a wide set of different areas that can be positively impacted by space technologies and space-based data. Despite this positive impact, space applications are often not acknowledged by end-users, making it difficult for decision-makers to purposefully implement policies that foster the utilisation of space in such macro-area (Fig. 4.9).

When it comes to the "environment and resources" area, responses collected via the survey attributed the highest scores to the United States (4.0), followed by Australia (3.5)—which scored very high in general, and in space for agriculture and meteorology in particular—and a group of countries with scores higher than "3.0": Brazil (3.4), China and Korea (both 3.3), Canada and India (both 3.27), and Europe (3.2). Respondents evaluated the lowest, in a descending order, Japan (2.92), Russia (2.65), and Israel (2.2).

As for "infrastructure," respondents showcased a general agreement on "mobility" being the field where all the countries scored the best (average 2.95)—particularly evident in Australia (4.0 in "mobility," 3.0 in "Infrastructure"), or Russia (3.7 vis-à-vis a general 2.75). The predominance of the indicator "mobility" in the "infrastructure" areas is due to the maturity of PNT-data based applications, having become more mainstream also thanks to the ubiquity of GNSS-enabled portable devices as the

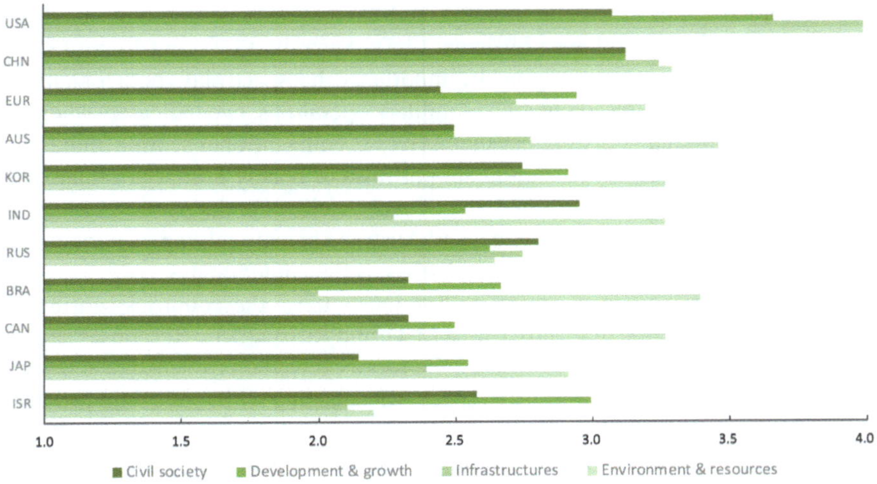

Reference data

	ISR	JAP	CAN	BRA	RUS	IND	KOR	AUS	EUR	CHN	USA
Environment & resources	2.2	2.9	3.3	3.4	2.6	3.3	3.3	3.5	3.2	3.3	4.0
Agriculture, Other natural resources, Meteorology, Environment, Climate change											
Infrastructures	2.1	2.4	2.2	2.0	2.7	2.3	2.2	2.8	2.7	3.2	4.0
Infrastructure management, Energy, Mobility											
Development & growth	3.0	2.5	2.5	2.7	2.6	2.5	2.9	2.5	2.9	3.1	3.7
Rural & urban development, Scientific & technological innovation, Industrial development, Market development & commercialisation											
Civil society	2.6	2.1	2.3	2.3	2.8	3.0	2.7	2.5	2.4	3.1	3.1
Health, Education, Entertainment & citizen services, National identity											
Socio-economic policies (*avrg*)	**2.5**	**2.5**	**2.6**	**2.6**	**2.7**	**2.8**	**2.8**	**2.8**	**2.8**	**3.2**	**3.7**

Fig. 4.9 Socio-economic policy

smartphone, or new space-based AIS applications, on which industries became more and more reliant over the last two decades (Fournier et al., 2018). Respondents consulted about the United States attributed to the country the top score for all the indicators in the area, followed by China (3.25) and Australia (2.8), while Israel and Brazil emerged as the lowest scoring countries (2.1 and 2.0).

Survey responses concerning the use of space for "development and growth" acknowledged the United States as the leading country—although without full marks (3.67)—followed by China (3.13), Israel (3.0), and Europe (2.9). The US was particularly rewarded in this area by its efforts to use space to develop markets and commercial activities– efforts that date back even to the Raegan Administration, with the attempts to commercialise US remote sensing programmes and Space Shuttle rides for commercial organisations (before the Challenger disaster) (Logsdon, 2021)—as well as to create an industrial base (3.67).

Finally, survey respondents considered China as the country using space the most in the "civil society" area (3.13). The country, for instance, quickly integrated space into its tools to foster a sense of national identity, developing its own "space culture" (Silk, 2021), with the launch of the already mentioned Tiangong Space Station being a significant moment for the country (Ref). China is followed by the United States (3.08) and India (2.96). Among the responses, it is worth highlighting the low scores attributed to Europe and Australia for what concerns the use of space to create/boost a national identity (respectively 1.6 and 1.3). This is something that might change in the future, with the EU launching its new Agency for the Space Programme (EUSPA, 2021) and the European Space Agency currently undergoing its most participated round of astronaut selections ever (Foust, 2021), or the Australian government leveraging the new enthusiasm coming from Australian space activities, also through the implementation of the Artemis Accords.

Overall, the average of the responses in the "socio-economic policies" macro-area shows that the highest scoring countries are the United States (3.69), followed by China (3.20), Korea (2.8) Australia (2.8), India (2.8), Europe (2.8), and Russia (2.71)—with the United States scoring significantly higher than the other countries (a 0.49 gap from the second country), being the country with a long history of programmes enabling the utilisation of space applications, such as that of Landsat (NASA, 2021c), or the GPS.

4.2.3 Building the Capacity Index and Matrix

Having described the results across the various areas of hard and soft capacity, it is now possible to aggregate results to obtain the total capacity for each of the considered countries, as explained in Chap. 3 of the book, with hard and soft capacity being the average of the scores that every country has obtained across the capacity macro-areas considered. Figure 4.10 shows the positioning of each one of the 11 actors included in the analysis within our capacity matrix.

The capacity matrix shows that only the United States, China, and Russia score higher than average (>3) in both capacity's subdimensions. Europe follows closely, but lags slightly behind the other three in the soft capacity dimension. All the remaining space actors seem inclined to integrate whatever space assets and skills they have (or may have access to) to support their national policies (thus developing their soft capacity) even though their hard capacity level is far below. The three major space powers plus Europe are far distant, with the US getting the highest scores in the group.

Figure 4.11 seems to suggests that the more balanced the relation between hard and soft capacity, the more a nation approaches the status of space power. However, it does not remain unnoticed that space powers tend to privilege hard over soft

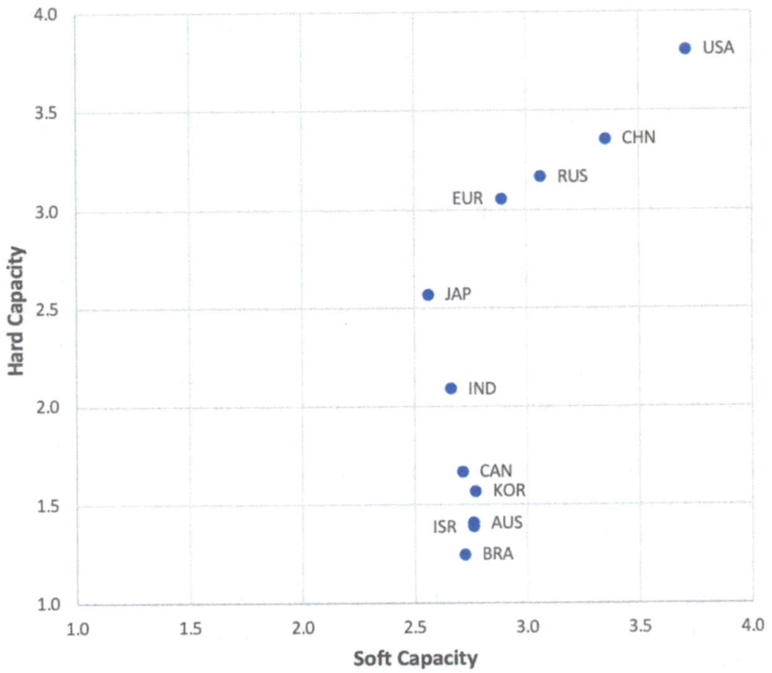

Fig. 4.10 Positioning of actors in the capacity matrix

capacity, while the exact opposite is true for the rest—with Japan in the almost perfectly balanced situation.

As for "hard capacity," not surprisingly, the United States leads the group of considered countries with a significant margin over the others. China ranks second behind the United States, having showcased an incredibly thriving space ecosystem led by bold ambitions. Russia comes third, immediately followed by Europe as single actor. Japan and India complete the group of countries scoring higher than 2, while Australia, Brazil, Canada, Israel, and South Korea all showcase relatively low hard capacity. Their relatively lower score can be explained by the fact that these countries have never fully engaged space across the full spectrum of applications and activities. The situation, however, might soon change given new impetus towards the development of the national space sector in some of these countries.

When looking at "soft capacity," the picture is somewhat different. The United States, China, and Russia still represent the leading countries, but Europe does not excel in soft capacity, obtaining scores similar to that of Israel, Canada Brazil, and Australia.

When calculating the "capacity index" for the countries, using the method explained in Chap. 3, we have the United States leading the group of countries followed by China, and then Russia. Once considered the leading space power due

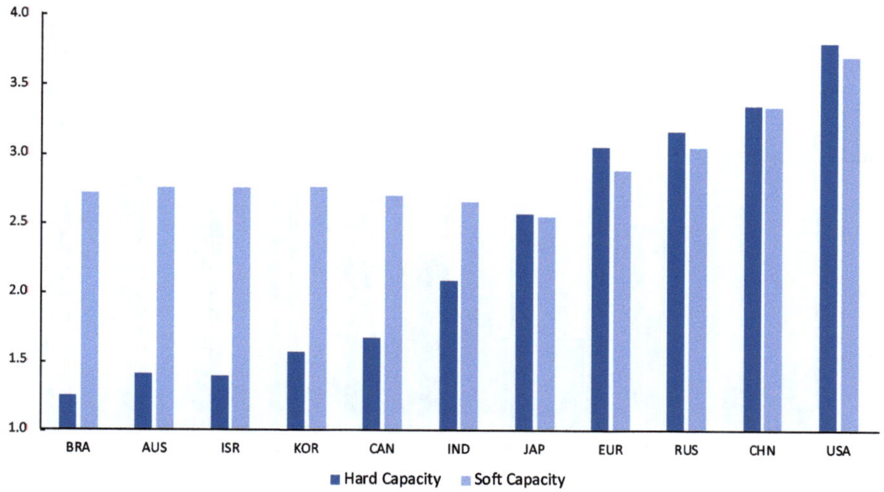

Reference data

	BRA	AUS	ISR	KOR	CAN	IND	JAP	EUR	RUS	CHN	USA
Hard Capacity	1.2	1.4	1.4	1.6	1.7	2.1	2.6	3.1	3.2	3.4	3.8
Soft Capacity	2.7	2.8	2.8	2.8	2.7	2.7	2.6	2.9	3.1	3.3	3.7
Capacity Index (*avrg*)	**2.0**	**2.1**	**2.1**	**2.2**	**2.2**	**2.4**	**2.6**	**3.0**	**3.2**	**3.4**	**3.8**

Fig. 4.11 Hard and soft capacity compared

to Soviet feats, Russia has been caught up by China in the last decade of space activities. Europe comes fourth, followed, by Japan and India, two other very active space actors.

4.3 Measuring and Comparing Autonomy

For the purpose of this work, "autonomy" denotes a state's ability to formulate space-related interests of its own and define national space strategies independent from or against the will of divergent interests (both foreign and domestic interests). Autonomy comprises a "technical" and a "political" sub-dimension. We refer to technical autonomy as "hard autonomy" and to political autonomy as "soft autonomy." In the next pages, we provide a detailed account of each of these two sub-dimensions.

4.3.1 Measuring Hard (Technical) Autonomy

Technical autonomy designates the state's ability to access and operate in space without relying on external sources of supply. It is articulated in three macro-areas: "production phase," covering space hardware and infrastructure, "operations phase," encompassing launch operations, satellite operations, and crewed operations, and "exploitation phase," which refers to data acquisition and service provision. Technical

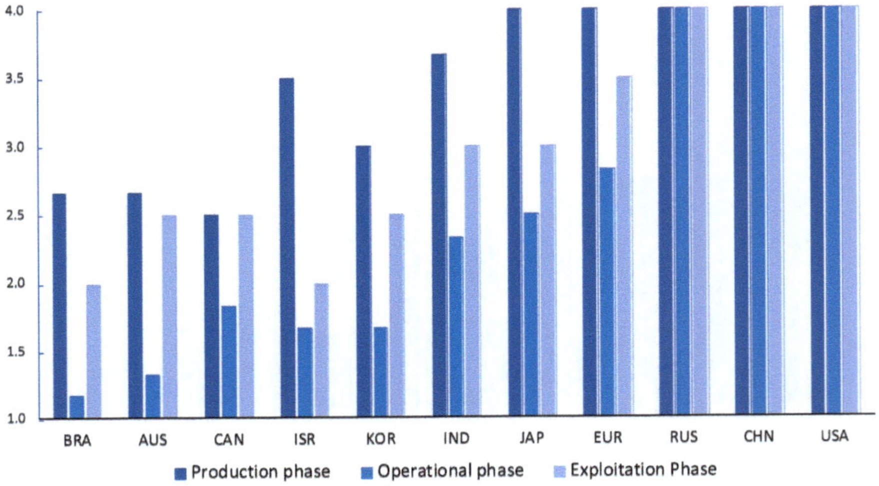

Reference data

	BRA	AUS	CAN	ISR	KOR	IND	JAP	EUR	RUS	CHN	USA
Production phase	2.7	2.7	2.5	3.5	3.0	3.7	4.0	4.0	4.0	4.0	4.0
Operational phase	1.2	1.3	1.8	1.7	1.7	2.3	2.5	2.8	4.0	4.0	4.0
Exploitation Phase	2.0	2.5	2.5	2.0	2.5	3.0	3.0	3.5	4.0	4.0	4.0
Hard Autonomy *(avrg)*	**2.0**	**2.2**	**2.3**	**2.4**	**2.4**	**3.0**	**3.2**	**3.4**	**4.0**	**4.0**	**4.0**

Fig. 4.12 Inside hard autonomy

autonomy comprises 13 indicators and builds on quantitative data directly collected by the research team. As for other sub-dimensions, scores run on a scale ranging from 1 (lowest) to 4 (highest) (Fig. 4.12).

Interestingly, the Unites States, China, and Russia all score the maximum in all three macro-areas. Europe and Japan also score the maximum for the production phase, but lag behind the three major powers when it comes to the operational the exploitation phases. While virtually all countries considered present reasonably high scores in the production phase, their performance varies widely when it comes to the operational phase.

4.3.1.1 Hard Autonomy: Production Phase

The "production phase" macro-area, which covers the technologies required for the development and manufacturing of space assets, including components, materials, and processes, is segmented in two main areas. The first is the production of space hardware, which includes the indigenous manufacturing of launch vehicles, satellite systems, modules, capsules, probes, and their sub-systems. The second concerns the ground infrastructure that enables the production and proper functioning of all these assets, primarily the assembly, integration and testing facilities and the broader ground segment (see Sect. 3.3.1 in Chap. 3). The overall results are displayed in Fig. 4.13.

For what concerns the area "space hardware," the United States, China and Russia, as well as Europe and Japan all score the highest (4.0), followed by India (3.33) slightly penalised by the entry "Module, capsules, and probes," given the country's relatively new attempts in crewed spaceflight. Israel also scores high (3.0), mainly thanks to the country's launch vehicles programme. When it comes to satellites owned and manufactured in the country, Australia and Brazil score the same (1.33), penalised by the lack of an active local launch capability programme in the countries during the period from 2011 to 2020, as well as by the countries' lack of participation in/conduction of space exploration missions. Canada scores the lowest, particularly penalised by the low ratio of locally manufactured satellites compared to those owned and operated by Canadian actors.

As for "infrastructure," all analysed nations show a perfect autonomy score (4.0) as for the two indicators concerning "Ground segment infrastructures" and "Assembly, integration, and test facilities," respectively. To conduct space activities, actors have to equip themselves with such kind of infrastructure, therefore it is not surprising that

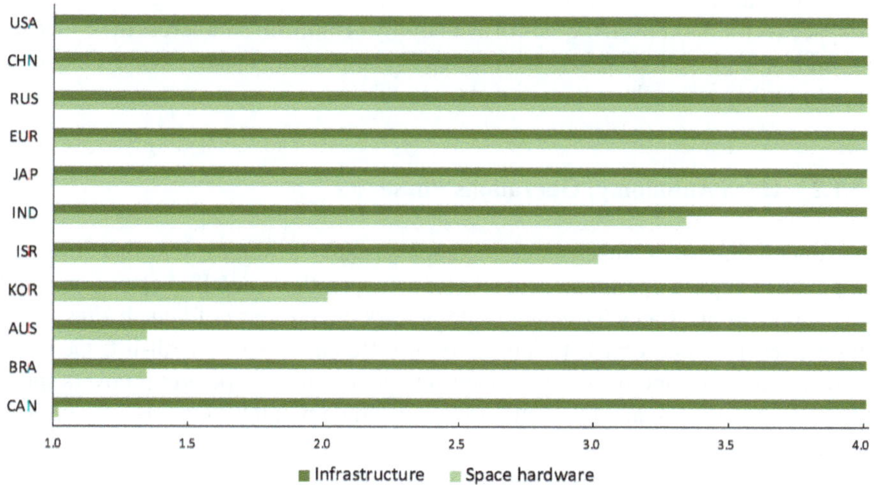

Reference data

	CAN	BRA	AUS	KOR	ISR	IND	JAP	EUR	RUS	CHN	USA
Space hardware Launch vehicles, Satellite systems, Modules, capsules, and probes	1.0	1.3	1.3	2.0	3.0	3.3	4.0	4.0	4.0	4.0	4.0
Infrastructure Ground segment, Assembly, integration and testing facilities	4.0	4.0	4.0	4.0	4.0	4.0	4.0	4.0	4.0	4.0	4.0
Production phase *(avrg)*	**2.5**	**2. 7**	**2.7**	**3.0**	**3.5**	**3.7**	**4.0**	**4.0**	**4.0**	**4.0**	**4.0**

Fig. 4.13 Production phase

the considered countries have developed adequate infrastructure to serve their spacefaring needs, from the governmental down to the "amateur" level of space activities. Also due to the relatively widespread use of radio technology connected to terrestrial communications, the countries score high for what concerns ground segment technology. As this infrastructure has been developed to match the national needs, the functionality and scale of these surely vary from country to county. Nevertheless, such infrastructure will have to evolve accordingly with the growing ambitions and scale of activities conducted by each country, which will see their autonomy hampered if they don't develop infrastructure at the same enabling rate as that of space activities to be supported by them.

This macro-area is therefore characterised by an overall high average score (3.45), positively impacted by the "infrastructure" area. Notably, traditional "spacefaring" countries such as China, Europe, Japan, Russia, and the United States receive the top mark (4.0), followed by India (3.67), Israel (3.5) and Korea (3.0), leaving the remaining countries below the 3.0 threshold due to a lower autonomy when it comes to building space hardware.

Overall, for the production phase, real technological autonomy would need to be verified down to the EEE micro-component elements utilised when building both space hardware and infrastructure. This is, however, an aspect which goes far beyond the space sector. Moreover, the lack of availability of uniform data across the countries as well as the globalised value chain for EEE components makes such evaluation a topic to further research on its own right.

4.3.1.2 Hard Autonomy: Operations Phase

The "operations phase" macro-area encompasses three different areas that cover the main types of space operations, namely launch operations, satellite operations, and crewed operations. Launch operations comprises the presence of launch sites in the national territory and access to launchers of different classes (small-lift, medium-lift, heavy-lift and superheavy-lift launch vehicles). Satellite operations covers space situational awareness, and telemetry, tracking and command functions. The crewed operations area encompasses astronaut training and human-rated operations. The overall results of this macro-area are shown in Fig. 4.14.

In the "launch operations" area, the United States, Russia, and China all obtain the highest score (4.0), followed by Europe and Japan (3.5 each), both countries penalised in autonomy in the indicator "Orbital access," due to the countries' lack of autonomy across the full array of launch vehicle classes—with the Europeans relying on Russian Soyuz vehicles for medium-class payloads, and the Japanese not having autonomous access to heavy or super-heavy launch vehicles. India follows (3.0), together with Israel (2.5). Australia, Brazil, and Canada receive the minimum score (1.0) due to the countries' lack of autonomy for what concerns both active launch sites and launch vehicles.

The "satellite operations" area is strongly defined by the capability to operate one's own spacecrafts fully autonomously. The reliance on foreign space domain

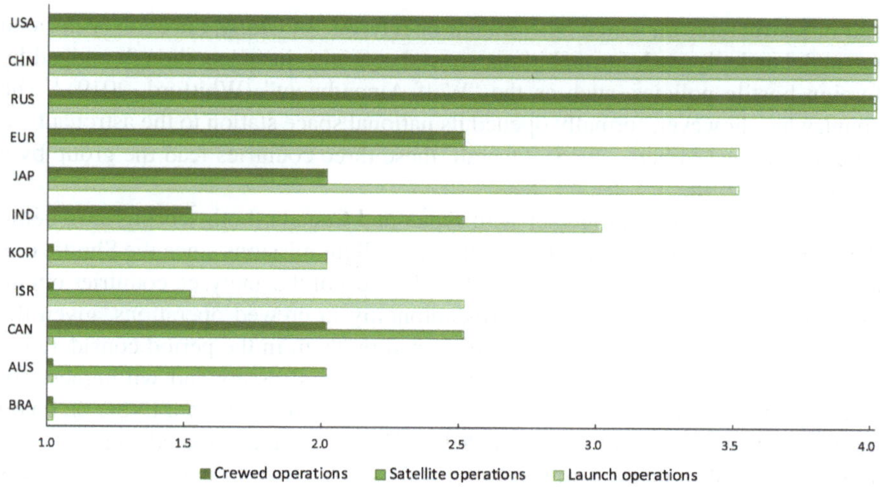

Reference data

	BRA	AUS	CAN	ISR	KOR	IND	JAP	EUR	RUS	CHN	USA
Launch operations Launch sites, Orbital access	1.0	1.0	1.0	2.5	2.0	3.0	3.5	3.5	4.0	4.0	4.0
Satellite operations Space domain awareness, Telemetry, tracking & command	1.5	2.0	2.5	1.5	2.0	2.5	2.0	2.5	4.0	4.0	4.0
Crewed operations Astronaut training, Human-rated operations	1.0	1.0	2.0	1.0	1.0	1.5	2.0	2.5	4.0	4.0	4.0
Operations phase (avrg)	**1.2**	**1.3**	**1.8**	**1.7**	**1.7**	**2.3**	**2.5**	**2.8**	**4.0**	**4.0**	**4.0**

Fig. 4.14 Operations phase

awareness capabilities and international sharing schemes impacts negatively almost all countries, with only the United States, China, and Russia obtaining the maximum score for "space situational awareness." These three countries score 4.0 overall, followed by Europe, India, and Canada (2.5), with Brazil and Israel scoring the lowest (1.5), particularly penalised by the dependency on foreign SSA capabilities and data.

Finally, in the "crewed operations" area, there is a clear connection between capacity and autonomy: countries with human spaceflight programmes and capacity are also those with the highest degree of autonomy in this regard. As the global space actors shifts towards more commercially driven human spaceflight activities, more countries will be able to simply acquire capacity by purchasing human spaceflight hardware and services, and this may likely create a decoupling of the capacity and autonomy dimensions for what concerns human spaceflight. Following this and the fact that, today, human spaceflight is inherently international in nature with most countries having to rely on, or cooperate with, third parties to conduct crewed activities, it comes as no surprise that the United States, Russia, and China are the countries scoring the highest (4.0)—with the first two countries having the most experience and infrastructures on which other countries rely. As for China, the country decided to singlehandedly push for ambitious crewed activities in space, though nudged by foreign hostile policies, such as the "Wolf Amendment" (Whitford, 2019).[1] The country has, however, formally opened its national space station to the astronauts of other countries (Aliberti, 2015). Overall, these three countries lead the group by a significant margin.

Europe follows (2.5), together with Japan and Canada (both scoring 2.0), being these historical contributors to US human spaceflight missions since the Shuttle era, and for this "preferred" partners of NASA. The rest of the analysed countries obtain the minimum score for what concerns autonomy in crewed operations, given the lack of human spaceflight activities conducted by them in the period considered—and hence the total reliance on international partners' assets and willingness for conducting some crewed activities (e.g., having a national astronaut flying on board the ISS as it happened for the UAE in 2019). The lack of human rated launchers and capsules naturally makes it impossible for these countries to obtain scores in this area.

Overall, for what concerns the "operational phase" macro-area, the United States, Russia, and China achieve top scores, with a significant gap between these and the rest of the considered countries. Europe (2.8), Japan (2.5), and India (2.3) follow with scores higher than 2.0. Brazil scores the lowest (1.2), particularly impacted by the complete lack of autonomy in both the area of "launch operations" and that of "crewed operations."

4.3.1.3 Hard Autonomy: Exploitation Phase

The third hard autonomy macro-area "exploitation phase," encompasses the main types of activities that need to be performed to effectively exploit satellite systems once in orbit, namely: data acquisition and service provision. Data acquisition assesses the autonomous access to satellite data, while service provision, the autonomous provision of satellite data-related services (Fig. 4.15).

As for the "exploitation phase" macro-area, the United States, China, and Russia showcase the highest autonomy (4.0), followed by Europe, India and Japan (3.0). All other actors score lower than 3.0.

Looking specifically in the areas considered, it is clear that autonomous "data acquisition" results difficult and impairing for many of the countries under review, as this is strongly linked to owning sovereign space assets which can provide that data. United States, China, and Russia score the highest (4.0), followed by Europe and India (3.0), and then Japan (2.0)—this last country particularly penalised by the

[1] Named after the former US Representative Frank Wolf, the Wolf Amendment is a clause passed by the United States Congress in the 2011 Appropriation Act prohibiting NASA and the White House Office for Science and Technology Policy from using government funds "to develop, design, plan, promulgate, implement, or execute a bilateral policy, program, order, or contract of any kind to participate, collaborate, or coordinate bilaterally in any way with China or any Chinese-owned company" (US Congress, 2011).

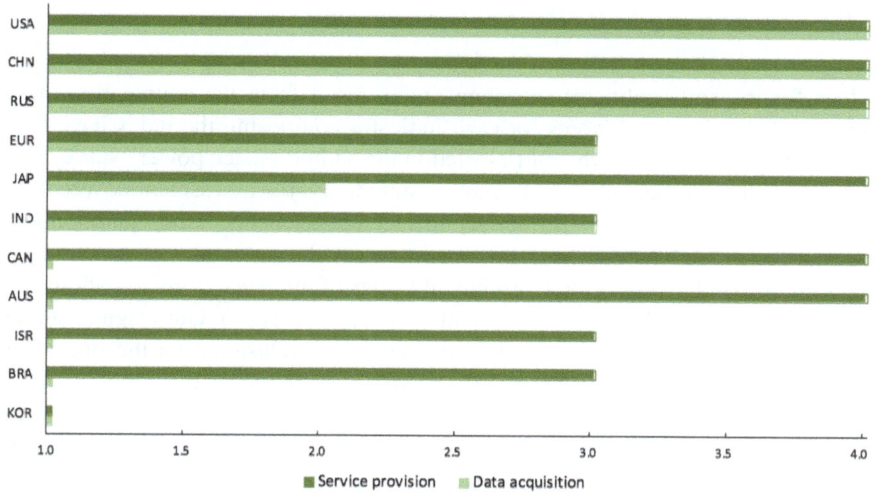

Reference data

	KOR	BRA	ISR	AUS	CAN	IND	JAP	EUR	RUS	CHN	USA
Data acquisition Autonomous access to satellite data	1.0	1.0	1.0	1.0	1.0	3.0	2.0	3.0	4.0	4.0	4.0
Service provision Autonomous provision of satellite data-related services	1.0	3.0	3.0	4.0	4.0	3.0	4.0	3.0	4.0	4.0	4.0
Exploitation Phase *(avrg)*	**1.0**	**2.0**	**2.0**	**2.5**	**2.5**	**3.0**	**3.0**	**3.0**	**4.0**	**4.0**	**4.0**

Fig. 4.15 Exploitation phase

reliance on foreign-built satcom satellites compared to its spacefaring counterparts. The remaining countries score all the minimum, given the low number of indigenous satellites thanks to which they could autonomously acquire data across the full spectrum of space applications.

On the other hand, for what concerns "service provision," all countries score fairly high, with the average score being 3.36, and with only Korea, Brazil, India, and Israel not receiving top mark, penalised by the lack of autonomous provision of SSA-related services. This overall result can be mainly explained with the fact that once a country has access to data—be it acquired autonomously or purchased/obtained through third-party means—it is easy to ensure a basic service provision, as almost any country likely has governmental agencies, academic institutions, or private companies with the ability to elaborate and deliver satellite products as a service.

4.3.2 Measuring Soft (Political) Autonomy

Values for the soft (political) autonomy index result from the aggregation of 18 expert-coded indicators. Scores for each indicator composing the index have been attributed by respondents who participated in the ad hoc "space power" survey.

Consistent with the conceptualisation of autonomy, the assessment covered two major macro-areas, namely "autonomy in external decisions" and "autonomy in internal decisions." Within these two macro-areas, the autonomy of space actors has been assessed vis-à-vis three prominent agents of influence: foreign nations, the domestic military, and domestic corporations (state or private). Our original survey data allows us thus to reach two different sets of conclusions: on the one hand, it allows us to identify *who* limits the autonomy of the state (i.e., the agents of influence); on the other, it allows for identifying *in which domains* the autonomy of the state is limited (i.e., the six areas). The distribution of the scores resulting from our survey leads to some tentative conclusions about the ability of space actors (states) to autonomously formulate *their own* policies. Figure 4.16 shows the overall results of the soft autonomy sub-dimension.

Within the soft (political) autonomy sub-dimension, we begin to see a different and apparently counterintuitive outlook. The most striking observation is that highly capable actors like Europe and the United States obtain very low results and full-fledged powerhouses like China and Russia do not perform well either. Europe, in particular, is the actor showing the lowest level of autonomy within this sub-dimension. This in part can be explained by the fact that Europe is not a nation-state in a Westphalian sense, nor is it governed by a single source of authority. However, the reasons behind this phenomenon are complex and should be understood within the wider decision-making processes of the countries under analysis. This includes considering their relationships with influential actors, both internal and external to the polity.

In order to better explain these results, we need to look inside the constituent macro-areas of the soft autonomy dimension.

4.3.2.1 Soft Autonomy: External Decisions

The "external decisions" macro-area assesses the participation of the analysed countries in the international space arena, mainly, though not only, in the context of multilateral organisations and fora, such as UNCOPUOS, the CD, ITU, ISO, IADC, ICG, etc. This macro-area notably encompasses three areas: "joining space-related treaties and organisations," "acting within space-related multilateral organisations," and "complying with space-related international arrangements." More broadly, this macro-area aims to assess whether and to what extent selected countries are able to autonomously define their space-related interests when it comes to foreign policy matters. As already discussed in Chap. 3, this ability can be variously limited by

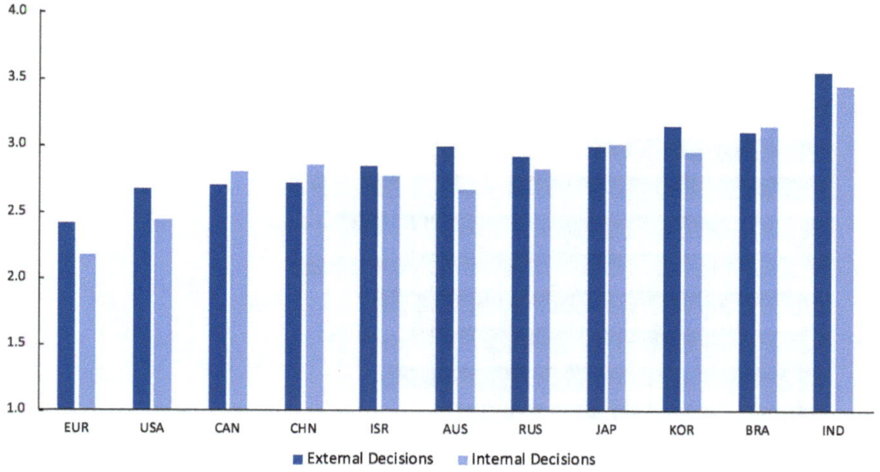

Reference data

	EUR	USA	CAN	CHN	ISR	AUS	RUS	JAP	KOR	BRA	IND
External Decisions* 'Joining,' 'Acting,' and 'Complying' in the international arena	2.4	2.7	2.7	2.7	2.8	3.0	2.9	3.0	3.1	3.1	3.6
Internal Decisions* Formulating space policy, Designing space programme, Choosing partners	2.2	2.4	2.8	2.9	2.8	2.7	2.8	3.0	3.0	3.1	3.5
Total Soft Autonomy *(avrg)*	**2.3**	**2.6**	**2.8**	**2.8**	**2.8**	**2.9**	**2.9**	**3.0**	**3.1**	**3.1**	**3.6**

* *Decision making autonomy from foreign nations, domestic military and corporations (both state and privately owned)*

Fig. 4.16 Inside soft autonomy

agents of influence, operating from both inside and outside the polity. The results of the "external decision" macro-area are shown in Fig. 4.17.

With regard to the area "joining space-related treaties and organisations," countries tend to cluster around mid-to-high values: the distribution of the scores ranges from a minimum of 2.40 to a maximum of 3.39. Surprisingly, the United States and China score among the lowest. Their relatively low score is primarily owed to the considerable influence of the national military (see further). Both Russia and the United States obtain very high scores on the first area indicator, which aims to assess the autonomy from foreign nations when it comes to the decision to join space-related bilateral and/or multilateral arrangements: this leads to the idea that both countries are very resistant to foreign pressures, and arguably even "dominant" in the international space-arena, meaning that they seek to influence, coerce, or buy other countries' decision-making on space-related issues. However, they both lack complete autonomy from various domestic-based actors. Indeed, the United States is apparently subject to significant influence by the national military (1.67) and domestic corporations (2.33).

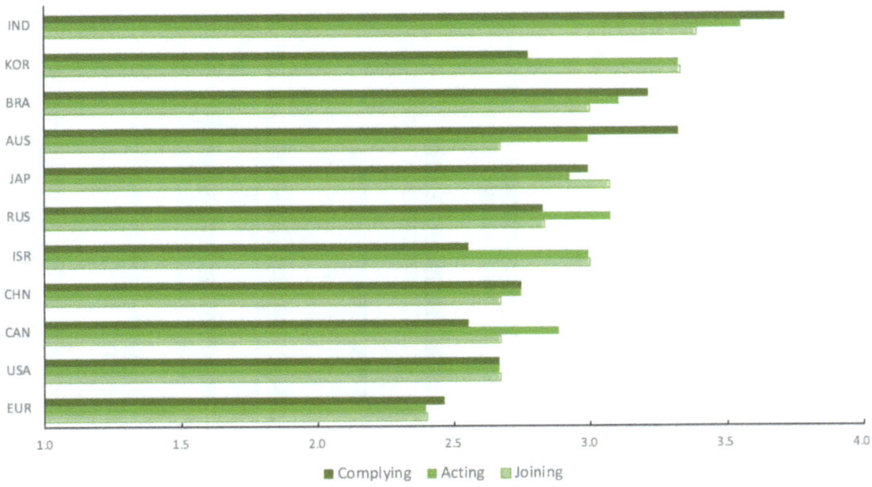

Reference data

	EUR	USA	CAN	CHN	ISR	RUS	JAP	AUS	BRA	KOR	IND
Joining space-related treaties and organisations*	2.4	2.7	2.7	2.7	3.0	2.8	3.1	2.7	3.0	3.3	3.4
Acting within space-related multilateral settings*	2.4	2.7	3.0	2.7	3.0	3.1	2.9	3.0	3.1	3.3	3.6
Complying with space-related international law*	2.5	2.7	2.6	2.7	2.6	2.8	3.0	3.3	3.2	2.8	3.7
Total External Autonomy by typology of decisions *(avrg)*	**2.4**	**2.7**	**2.8**	**2.7**	**2.9**	**2.9**	**3.0**	**3.0**	**3.1**	**3.1**	**3.6**

* *Decision making autonomy from foreign nations, domestic military and corporations (both state and privately owned)*

Fig. 4.17 External autonomy by typology of decisions

India leads the ranking with a score of 3.39, with South Korea finishing a close second (3.33) and demonstrating a nearly complete autonomy from national private actors when it comes to joining space-related international arrangements.

Similarly, the distribution of the scores within the area "acting within space-related multilateral settings" is concentrated around mid-to-high values. The gap between the leading space actors, India (3.6), Korea (3.3), and Brazil and Russia (3.1) is just 0.5 points. This group is followed closely by Australia, Israel and Canada (3.0).

When it comes to the area "complying with space-related international arrangements," India leads the ranking with a remarkable 3.72 score, marking the highest value in the overall political autonomy survey. Particularly, India scores one point or more higher than China, Canada, Israel, the United States, and Europe, which finishes last in the area ranking. Yet, while Canada, Europe and Israel's main weakness concerns dependence to foreign nations (2.33), the United States appears more vulnerable to the influence of the military-industrial complex.

Overall, in the "external autonomy" macro-area, India (3.57), South Korea and Brazil (3.1) stand out as the most autonomous space actors. This means that this group proves resilient to both domestic and foreign interests when it comes to defining and pursuing their space-related interests in the international space arena. Six actors slightly fall below the "3.00" score threshold: Europe, the United States, Canada, China, Israel, and Russia. The distribution of the scores provides evidence for the hypothesis formulated above: unlike other countries, Russia, the United States, and to a minor extent China—widely regarded as the most prominent space actors of the twenty-first century—are virtually immune to foreign pressure. Anyway, they are subject to intrusive influence by national actors, especially the military.

This leads to a somehow counterintuitive observation: while these actors (particularly the United States) are believed to exert a dominant role in the international space arena and to pose strict limits on the external autonomy of their partners/clients, their ability to formulate autonomous political preferences is often "captured" by powerful agents of influence such as big corporations and the military. This observation can be further understood and evidenced by looking at the external autonomy by typology of agents of influence (see Fig. 4.18).

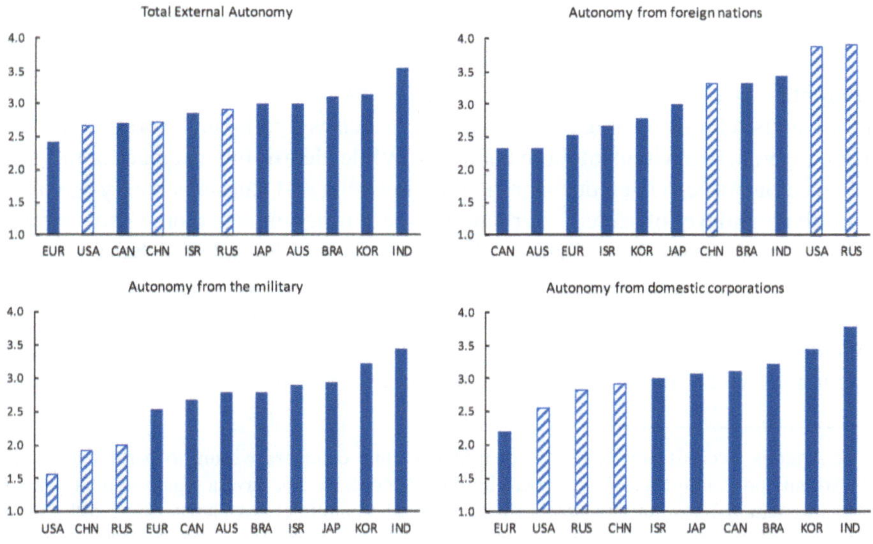

Reference data

	EUR	USA	CAN	CHN	ISR	RUS	JAP	AUS	BRA	KOR	IND
Total External Autonomy	**2.4**	**2.7**	**2.7**	**2.7**	**2.9**	**2.9**	**3.0**	**3.0**	**3.1**	**3.1**	**3.6**
from foreign nations	2.5	3.9	2.3	3.3	2.7	3.9	3.0	2.3	3.3	2.8	3.4
from the military	2.5	1.6	2.7	1.9	2.9	2.0	2.9	2.8	2.8	3.2	**3.4**
from domestic corporations	2.2	2.6	3.1	2.9	3.0	2.8	3.1	3.9	3.2	3.4	3.8

* *Decisions making autonomy in 'Joining,' 'Acting,' and 'Complying' in the international arena*

Fig. 4.18 External autonomy by typology of agents of influence

Starting with the "autonomy from foreign nations" that analysed actors hold when joining, acting and complying with space-related international arrangements, Canada, Australia and Europe figure as the most sensitive actors to the inducements or threats of third countries, particularly the United States. Fairly low scores, however, are also obtained by Israel, South Korea and Japan, which similarly demonstrate to be particularly sensitive about US preferences.

Although no significant cases of undue influence in space-related international settings have been substantiated, a notable and non-coincidental alignment is observable between the international stance of these countries and that of the US, especially when decisions regarding the political behaviour to be adopted in the international arena directly affect US interests and security concerns.

It is for instance no coincidence that the above countries are notable signatories of both the US-led Artemis Accords[2] and the ASAT moratorium.[3]

In terms of the "autonomy from foreign nations" ranking, both Russia and the United States exhibit an almost complete autonomy from foreign nations when it comes to deciding to join space-related bilateral and/or multilateral arrangements, acting within space-related multilateral settings, or complying with space-related international arrangements.

When looking at "autonomy from the military," however, we see a completely different picture. In this case the United States, Russia and China emerge as the least autonomous actors, meaning that the military exercises significant levels of influence on the external behaviours of these countries. While identifying instances of military influence on space-related decision-making in China and Russia is barely possible, given the high level of secrecy surrounding space programmes, there is no shortage of research showing the influence of US military on the political behaviour the US government on major space governance issues, including those pertaining to the civilian sphere. For instance, in his research about the role of the military on both domestic and international decisions related to space safety and sustainability, Verspieren (2020) demonstrated that in the US the military enjoys an *absolute*

[2] The Artemis Accords are 7-page set of principles that countries willing to participate in the US Artemis programme accept to adhere by. The Accords do not have a legally binding nature under international law but represent a strong political commitment by signatories (ESPI, 2020). More than 20 countries signed the Accords. While the Accords are, title-wise, associated with the Artemis programme, principles and provisions therein reach beyond Artemis-related lunar presence and apply to participating countries' activities on "Moon, Mars, comets and asteroids." Obviously, a non-signature of the Accords does not prevent other countries from engaging in space exploration. However, if a country wants to collaborate with the USA, it is reasonable to expect that NASA would request signature to the Artemis Accords. This conditioning of cooperation through acceptance of space governance principles underscores that the Artemis Accords can be seen as a US diplomatic tool for pushing US preferences on international space governance matters (ESPI, 2020).

[3] In April 2022, the US became the first country to declare a self-imposed ban on ASAT tests. Immediately after, the US government started looking at ways to "multilateralise" the test ban, encouraging other countries, its allies in particular, countries to go on record regarding their support (Foust, 2022). Canada joined the moratorium in May 2022, New Zealand in July 2022, Germany and Japan in September 2022, while South Korea and the United Kingdom in October of the same year. Other allies have expressed support but have not joined the moratorium.

proscriptive influence on major international governance issues, meaning that "if the military opposes a position on space safety and sustainability, then this position is perceived as unacceptable by other agencies involved in domestic decision-making." The research shows that US military also enjoys *strong prescriptive influence*, meaning that "if the military supports a position on space safety and sustainability, then this position is perceived very favourably by other agencies involved in domestic decision-making" (Verspieren, 2020).

On the other side of the spectrum, India proves to be the most autonomous from military influence in external decision-making processes. One factor that may contribute to explaining this is the way India's foreign policy establishment operates, in particular the structure of the Indian Foreign Service (IFS), which produces a decision-making process that is highly individualistic (Aliberti, 2018). All three major governmental bodies responsible for making India's foreign policy—namely, the Prime Minister Office, the National Security Office and the Ministry of External Affairs—are filled in their top positions by officers of the Indian Foreign Service (IFS). As Chatterjee Miller contends: "since foreign service officers are considered the crème de la crème of India and undergo extensive training, they are each seen as capable of assuming vast authority. What is more, the service's exclusive admissions policies mean that a tiny cadre of officers must take large portfolios of responsibility. In addition to their advisory role, they have significant leeway in crafting policy. This autonomy, in turn, means that New Delhi does very little collective thinking about its long-term foreign policy goals, since most of the strategic planning that takes place within the government, happens on an individual level" (Chatterjee Miller, 2013).

Similarly, when looking at the "autonomy from corporations" ranking, scores once again identify India as the most autonomous player when it comes to political decisions adopted in the international space arena. Brazil, South Korea, Canada and Japan follow at a distance. On the contrary, what is most striking is the very low scores obtained by leading space actors such as Europe, the United States, Russia and China. What these actors have in common is the presence of strong and influential space corporations—be them private or state-owned—that are able to exercise significant influence in external decision-making processes.

In the case of Europe and the United States, which were assessed as the least autonomous actors most susceptible to the intrusion of industry's particular interests, influence is often exercised through powerful trade associations providing representation of the leading domestic satellite and launch providers, operators, manufacturers, and suppliers such as the Paris-based ASD-Eurospace and the US-based Satellite Industry Association (SIA). An illustrative example is offered by the proposal led by the US government and backed by US corporations for an ISO Space Traffic Coordination and Management (STCM) standard, where most of the requirements were derived from the Open Architecture Data Repository (OADR). This architecture was quite different from the current EU SSA/SST architecture, and this explains why European space industries lobbied decision-makers of European member states to reject the proposed standard, as its adoption would have led to a competitive advantage for US companies over European ones, negatively affecting their business models.

4.3.2.2 Soft Autonomy: Internal Decisions

The "internal decisions" macro-area focuses on the domestic space policies of the 11 selected space actors, expanding on the political decision-making processes leading to the definition and implementation of their national space interests, strategies, and objectives. It embraces three prominent areas: "formulating a national space policy," "defining a national space programme," and "choosing partners within the space domain." It aims to assess whether, and to what extent, countries are able to autonomously formulate their interests when it comes to domestic space policy, especially vis-à-vis other nations, the domestic military complex and corporations. The results are displayed in Fig. 4.19.

In the "formulating a national space policy" area, India and Brazil score the highest, followed closely by China and Israel. The distribution of the scores tends to the upper limit of the 1–4 range, with Japan (2.95) slightly falling below the 3-point threshold and the United States and Europe finishing distantly (respectively, 2.2 and 2.3). Importantly, the United States obtains the lowest possible score on the second

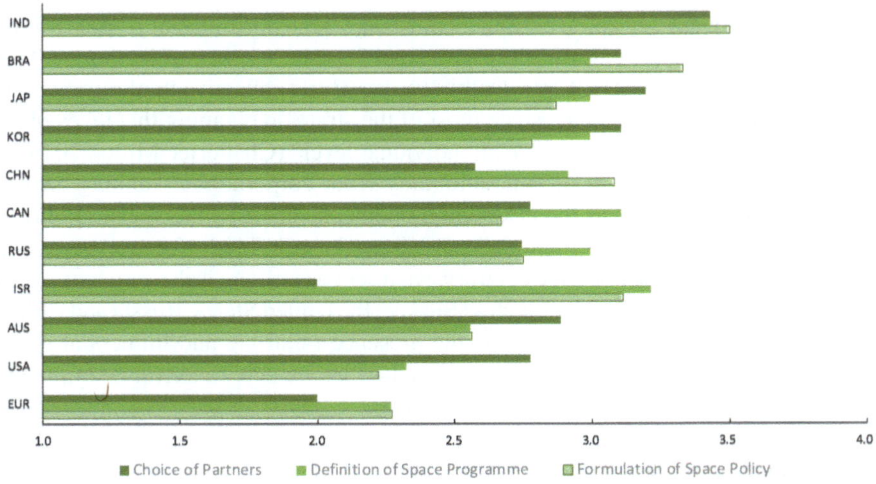

Reference data

	EUR	USA	AUS	ISR	RUS	CAN	CHN	KOR	JAP	BRA	IND
Formulating a national space policy*	2.3	2.2	2.6	3.1	2.7	2.7	3.1	2.8	2.9	3.3	3.5
Defining a national space programme*	2.3	2.3	2.6	3.2	3.0	3.1	2.9	3.0	3.0	3.0	3.4
Choosing partners within the space community*	2.0	2.8	2.9	2.0	2.7	2.8	2.6	3.1	3.2	3.1	3.4
Internal Autonomy by typology of decisions *(avrg)*	**2.2**	**2.4**	**2.7**	**2.8**	**2.8**	**2.9**	**2.9**	**3.0**	**3.0**	**3.1**	**3.4**

* *Decision making autonomy from foreign nations, domestic military and corporations (both state and privately owned)*

Fig. 4.19 Internal autonomy by typology of decisions

indicator (1/4), which aims to assess the autonomy vis-à-vis the military. In other words, according to prominent experts in the space field, the military exerts a nearly total influence on the decision-making process in the United States when it comes to formulating a national space policy.

India leads the ranking in the "defining a national space programme" area, scoring an almost perfect 3.4 and facing minor issues only when it comes to autonomy from the national military (3.17). India is followed by Israel (3.2) and Canada (3.1), with respect to autonomy vis-à-vis corporations as well as foreign countries. Both China (2.94) and Europe (2.3) receive a score lower than 3, despite facing very different constraints: while China shows greater dependence on the national military (2.00), Europe mainly lacks autonomy vis-à-vis economic powerhouses (2.67). Again, the United States (2.3) ranks last, tied with Europe, being subject to severe pressures from the military (1.33) and domestic corporations (2.00).

Finally, India (3.4) obtains the highest score in the "choosing partners within the space domain" area, followed closely by Japan (3.2), Brazil (3.1) and Korea (3.1). Israel and Europe finish last (2.1), with Canada and the United States (2.8) slightly falling under the 3-point score threshold. Worth to notice, when it comes to selecting partners for implementing the space policy agenda (e.g., technology or raw material suppliers), Russia shows a significant dependence on the national military (1.75): only Israel receives a lower score for the same indicator (1.67), while Australia scores the highest among the countries under consideration (3.67).

Overall, in the "internal decisions" macro-area, the distribution of the scores is significatively concentrated around the 3-point score threshold, with India (3.5) leading as the most autonomous actor. Brazil (3.1), Japan (3.0), and Korea (3.0) follow India, despite facing very different situations: Brazil is to some extent subject to pressure from the national military; Japan appears to be relatively vulnerable to its domestic industrial complex. The United States received the second worst score (2.4), only besting Europe (2.2). Again, while being virtually immune to foreign pressures, the Unites States lacks autonomy vis-à-vis both the military and domestic corporations.

Once again, results in the internal autonomy macro-area can be better explained and appreciated by looking at the survey results from the perspective of the agents of influence, rather than the specific issue-areas (see Fig. 4.20).

When looking at "autonomy from foreign nations," Australia, Canada and Europe once again figure as the actors most vulnerable to foreign invitations and/or pressures, particular those coming from their closest ally, the United States. This sensitivity proves to be closely interwoven to the security and defence dimensions permeating space activities, including civilian ones. A well-documented case is the interrupted cooperation between the European Union and China within the Galileo programme due to the strong pressure—and even threat of retaliation—from Washington (Aliberti, 2015; Casarini, 2009). As poignantly put by Johnson-Freese and Erickson (2006), in its broader space relations with Chinese space organisations, Europe has been generally left with the "dilemma of either not expanding cooperation with China, a restriction that it wants to avoid, or of risking the wrath of the USA, which it neither wants, nor most probably can afford."

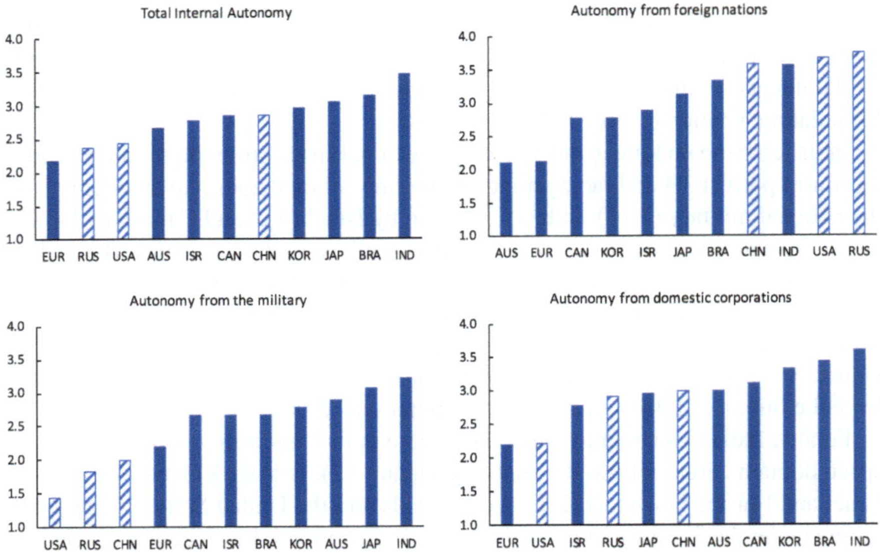

Reference data

	EUR	RUS	USA	AUS	ISR	CAN	CHN	KOR	JAP	BRA	IND
Total Internal Autonomy* *(avrg)*	**2.2**	**2.4**	**2.4**	**2.7**	**2.8**	**2.9**	**2.9**	**3.0**	**3.1**	**3.1**	**3.5**
from foreign nations	2.1	3.8	3.7	2.1	2.9	2.8	3.6	2.8	3.1	3.3	3.6
from the military	2.2	1.8	1.4	2.9	2.7	2.7	2.0	2.8	3.1	2.7	3.2
from domestic corporations	2.2	2.9	2.2	3.0	2.8	3.1	3.0	3.3	3.0	3.4	3.6

* *Decision making autonomy in Formulating a space policy, Designing a space programme, Choosing partners*

Fig. 4.20 Internal autonomy by typology of agents of influence

Besides this specific instance, it is worth noting that even countries scoring mid-to-high autonomy (namely, South Korea, Brazil, Japan and Israel) are reportedly permeable to foreign *coercion* and *inducements*. With regard to the formulation of national space policies, well documented, for instance, is the influence recurrently exercised by the United States on Japan to make its space policies and regulations more attuned to US interests (Aoki, 2009; Kallender, 2016; Zeng, 2004). With regard to the definition of national space programmes, particularly eloquent is that the development of domestic launch capabilities by South Korea and Brazil have been for a long time halted, through the recourse of diplomatic pressures and threat of retaliation, by the United States, which feared that orbital rocket programmes would accelerate missile proliferation (Moltz, 2012; Newberry, 2003).

Not surprisingly, on the other hand, Russia, China and the United States lead this specific ranking, demonstrating basically no vulnerability to foreign nations' influence in their policy-making processes. On the contrary, these countries are constantly portrayed as seeking to influence, coerce, or buy other countries' decision-making on space-related issues. The Prague Space Security Studies Institute (PSSI), for instance, has provided evidence of space sector capture practices by China and Russia in forging their international cooperation with third countries (Robinson, 2019). Interestingly, space sector capture has been defined by the PSSI as "a state actor's provision of space-related infrastructure/equipment, technology, launch and other services, education/training and financing ultimately designed to limit the freedom of action and independence of the recipient state's space sector, generally implemented on an incremental basis" (Robinson, 2019).

When looking at "autonomy from the military," instead, the scores provide a completely different picture. In this case it is the three leading space powers that emerge as the least autonomous actors. Indeed, the United States (1.44), Russia (1.83) and China (2.00) are reportedly subject to significant influence by their national military when it comes to formulating national space policies, defining programmes and choosing partners in the space arena. This is not surprising considering that space has historically been "a significant contributor to military power" (Swilley, 2011), thus raising high stakes for making national policies attuned to the interests of the military apparatuses. But in these countries the role of the military is seen as an overwhelming one. For instance, Chinese space programme has been recurrently described by Western analysts as being inherently military in nature, due the active involvement of the People Liberation Army (PLA) in the management and execution of all national space activities. Reports and testimonies produced by the US-China Economic and Security Review Commission for the US Congress[4] have repeatedly emphasised how the key infrastructural elements (like all launch and control facilities) are run and staffed by the military, and even purportedly civilian endeavours such as human spaceflight also see their direct involvement. While affirming that projects are run by the PLA should not automatically imply that they are ultimately decided on and controlled by the military,[5] the influence of the PLA in the Chinese decision-making processes is certainly far from marginal.

On the other side of the spectrum, the bar plots shows that India and Japan are the most resilient actors vis-à-vis the military. This can be explained by the fact that both Japan's and India's national space programmes have been characterised since their beginning by a strong civilian focus, rather than being driven by military interests. Historically, the opposite has been the case in the US, China and Europe (Aliberti, 2018).

In the case of India, until the last decade the military had minimal involvement in the country's space activities, with no access to dedicated imagery, satellite-based communications or data services. Instead, they had to rely on foreign assets such as purchasing imagery from US companies and the Israeli government, among other sources. The management of space activities was exclusively entrusted to the civilian ISRO (Aliberti, 2018). Behind the government's decision to firewall any military

---

[4] See for example the Hearing on "China in Space: A Strategic Competition?" held in Washington DC, United States in April 20,219 by US-China Economic and Security Review Commission (2019).

[5] As a matter of fact, "not only are core responsibilities shared with other leading stakeholders (e.g., SASTIND, the MOST and CAS), but key decisions on the implementation of space policies and the overall direction of the programme ultimately reside in the hands of the high-level decision makers of the Party" (Aliberti, 2015).

component from the management of the programme, there was the view that the Indian space programme would have been unnecessarily slowed if it had been a combined military and civil space programme, and that the cooperation with other states would have been impossible if India had possessed a military space programme.

In the case of Japan, the use of space technology by the military has been restricted for half a century due to a firm non-military interpretation of the principle of the peaceful use of outer space contained in the Outer Space Treaty (Aliberti & Hadley, 2020). This interpretation, sanctioned in 1969 by the Japanese Diet, barred the military from being involved or exploring opportunities in space, i.e., using space for reconnaissance, surveillance, communications and early warning purposes (Peoples, 2013). It also prevented commercial actors from developing space technology with an explicit military purpose (Suzuki, 2015).

Whereas the space policy posture of both India and Japan has witnessed—and is still witnessing—dramatic changes over the past decade, with military involvement in the sector gaining increasing traction, the military still has a very small influence on the overall formulation of space policies in both countries.

When looking at the "autonomy from corporations," the results once again show that the national space industry is no stranger to exert significant influence on the domestic space policy process, as demonstrated by the low average scores received by highly capable actors such as the United States, China, Russia and Japan. The bar plot more specifically suggests that countries possessing a more siseable commercial space industry tend to be less autonomous from corporations' interest when articulating internal space policy decisions and defining programmatic activities. It is in fact in Europe (2.20) and the United States (2.20) that domestic corporations stand out as the most authoritative agents of influence. Literature indeed shows (OECD, 2021; Mazzucato et al., 2016; Tugnoli & Wells, 2019) how public agencies in these countries have become increasingly reliant on private actors to accomplish their missions. Such reliance has undeniable benefits, as it enables space agencies and other public institutions to optimise resource allocation and re-focus on technological development and maturation of early-stage technologies and scientific research, while leaving mature technologies to the industry. Ultimately, increased industrial dynamism adds to the baskets of space capacities, both hard and soft, that allow states to assert themselves as space powers. At the same time, however, such reliance also comes with non-negligible costs for public actors. More specifically, greater industry dynamism is increasingly associated to growing levels of influence and hence to reduced levels of political autonomy.

On the other side of the spectrum, India appears to enjoy almost complete decision-making autonomy from its domestic corporations. The result is consistent with the fact that the involvement and capabilities of private industries in the Indian space programme remain rather marginal in comparison to the United States and Europe. As a matter of fact, in both the upstream and downstream segments of India's space industry, the role of the public actor is quasi-monopolistic, with ISRO and its subordinate entities acting as the main manufacturer/provider of space products and as the main operator for the commercialisation of space-related services (Aliberti, 2018). Whereas India's space sector can boast an industrial base of more than 500 small, medium and large enterprises and the production of many satellite systems is reported to be almost completely outsourced to industry, reality is that the current supply chain for space infrastructure development and exploitation continues to be centred around ISRO (Prasad, 2016a). Within the current industry participation model, most of the private industry in India remains involved in the supply of Tier-2/Tier-3 products and services for ISRO (mainly subsystems components, as well as equipment and engineering services). There are a handful of Tier-1 space companies (delivery of complete subsystems) in the industrial ecosystem, a role that still belongs to ISRO. The final assembly, integration and testing (AIT) of spacecraft and launch vehicles are also conducted by ISRO (Prasad, 2016b). Overall, this makes the influence of corporations in India's decision-making processes negligible.

4.3.3 Building the Autonomy Index and Matrix

Having obtainined the scores for both technical and political autonomy, it is now possible to aggregate them and calculate the total autonomy value for each of the selected countries, according to the framework described in Chap. 3.

An overview for the countries' autonomy in space activities emerges, with hard and soft sub-dimensions portraying two very different pictures.

Figure 4.21 shows the placement of all 11 countries included in this study within the autonomy matrix. The two dimensions, namely hard autonomy and soft autonomy, contribute in rather different ways to the positioning of our countries on the map. With the exception of Japan, there exists an inverse relation between the two dimensions for most countries. Moreover, for those that are *not* major space powers the relation tends to be: *the lower the level of hard autonomy, the higher the level of soft autonomy*. The opposite is true for the three major space powers (and Europe): *the higher the level of hard autonomy, the lower the level of soft autonomy* (Fig. 4.22).

As already discussed with reference to the case of human spaceflight and crewed operations—but arguably scalable to the whole space tech realm—while space activities continue to not transition to fully commercial solutions which any nation can simply purchase, it can be expected that hard capacity and technical autonomy will be strongly related. This is reflected in the scores for "hard (technical) autonomy," where the countries most capable of conducting space activities are also those with higher technical autonomy. United States, China, and Russia score the highest (4.0) followed by Europe (3.44), Japan (3.17), and India (3.0). The other countries are significantly behind, averaging 2.23.

A closer examination of "soft (political) autonomy" reveals a distinct contrast in the scores. As per the responses from the surveyed experts, countries that scored higher than 3.0 in this dimension include India, followed by Brazil, South Korea, and Japan. Remarkably, Canada, China, Israel, and Europe closely trail behind, while the United States receives the lowest political autonomy score (2.6). The United States' limited performance in this field, particularly with regard to domestic agents

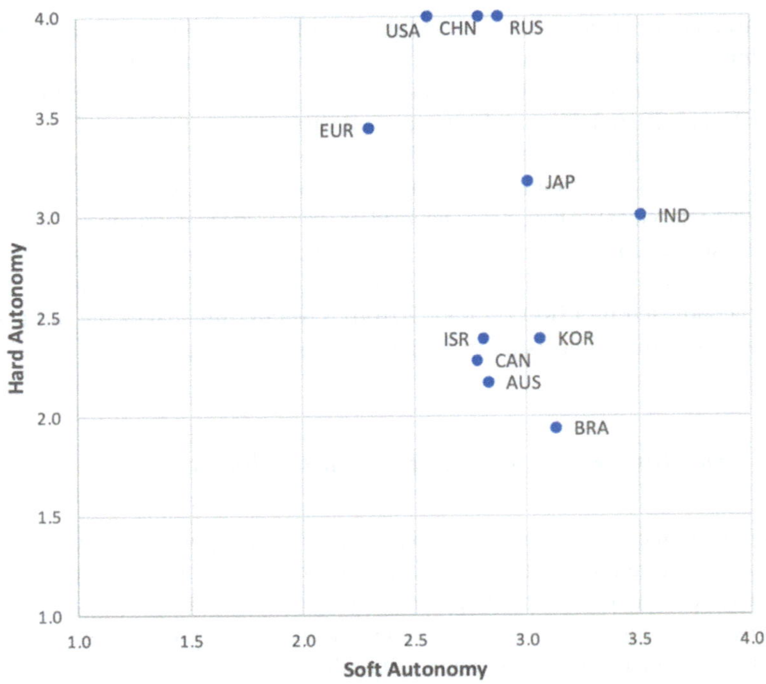

Fig. 4.21 Autonomy matrix

of influence, is consistent with the literature on lobbying and regulatory capture examined in chapter 2.2. It is also likely linked to the intricacies of the policy process expected in a democratic society. The sheer number of stakeholders involved in the decision-making cycle and the constant scrutiny of space programmes by the national community in the United States may contribute to explain this comparatively lower score.

When calculating the values for the "autonomy index," as explained in Chap. 3 and demonstrated in, Figure 4.22 Russia (3.4) and China (3.4) emerge as the two leading countries, followed by the United States (3.3), India (3.2), Japan (3.1), and Europe (2.9)— with all these countries short of India displaying a political autonomy score lower than the technical autonomy one. The opposite is true for the remaining countries, which are penalised by lower technical autonomy scores.

4.4 Measuring Spacepower and Identifying Space Powers

Figure 4.23 compares our autonomy and capacity indices across all 11 space actors included in this analysis. It clearly shows that the autonomy index generally exceeds the capacity one; only in the US and in Europe does the autonomy average fall below

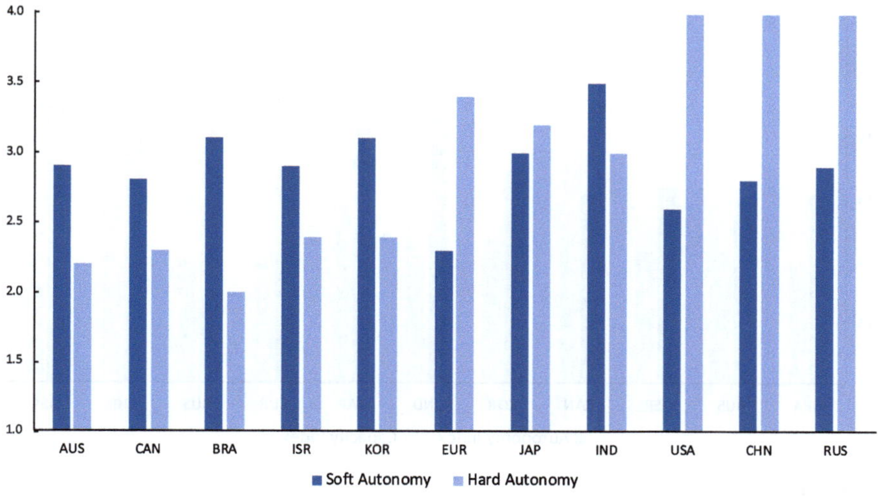

Reference data

	AUS	CAN	BRA	ISR	KOR	EUR	JAP	IND	USA	CHN	RUS
Soft Autonomy	2.9	2.8	3.1	2.9	3.1	2.3	3.0	3.5	2.6	2.8	2.9
Hard Autonomy	2.2	2.3	2.0	2.4	2.4	3.4	3.2	3.0	4.0	4.0	4.0
Autonomy Index *(average)*	**2.6**	**2.6**	**2.6**	**2.7**	**2.8**	**2.9**	**3.1**	**3.3**	**3.3**	**3.4**	**3.5**

Fig. 4.22 Hard and soft autonomy compared

the capacity average; China and Europe show the most balanced proportion between the two dimensions, with the former being slightly more autonomous than capable and the latter being slightly more capable than autonomous. The bulk of our space actors (i.e., Australia, Brazil, Canada, India, Israel, Japan, and Korea) can be defined "autonomous spacefaring nations," in the sense that their levels of autonomy are higher than their levels of capacity. India presents the worst balance between the two dimensions of all cases analysed, with high levels of autonomy and relatively low levels of capacity. Noticeably, Russia is the only high-performing space power that presents an unbalanced relation between autonomy and capacity, similar to the unbalanced patterns of the "autonomous spacefaring nations."

Based on scores featured in Fig. 4.23, space actors can be finally mapped within the spacepower matrix and comparatively assessed in relation to each other. In Fig. 4.24, we propose again the visualisation of the spacepower matrix with the relative positioning of the analysed players.

After having plotted the various space actors along the two axes, it becomes possible to analyse and comment upon their distribution in the spacepower taxonomy matrix proposed in Chap. 2.

Figure 4.25 shows that the 11 space actors considered in our analysis are clustered in 3 distinct groups. The first group includes the United States, China, and Russia, the three greatest space powers in the world. In comparison to the other space actors in our analysis—any other space actor in the world—these three countries are in a

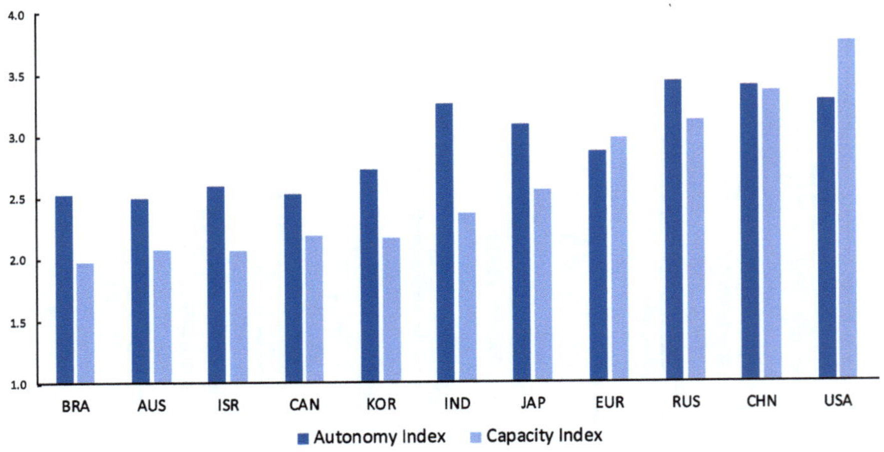

Reference data

	BRA	AUS	ISR	CAN	KOR	IND	JAP	EUR	RUS	CHN	USA
Autonomy Index	2.5	2.5	2.6	2.5	2.7	3.3	3.1	2.9	3.4	3.4	3.3
Capacity Index	2.0	2.1	2.1	2.2	2.2	2.4	2.6	3.0	3.1	3.4	3.8

Fig. 4.23 Capacity and autonomy compared

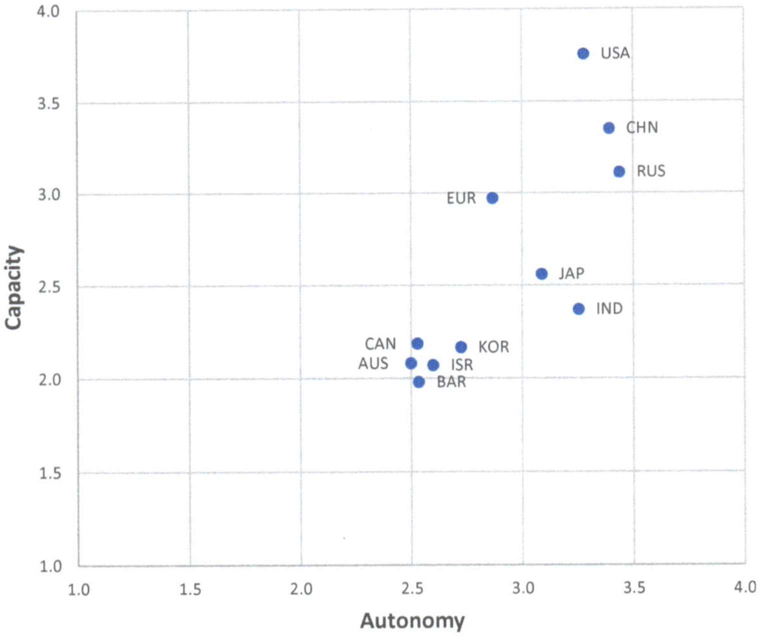

Fig. 4.24 Positioning of space actors in the spacepower matrix

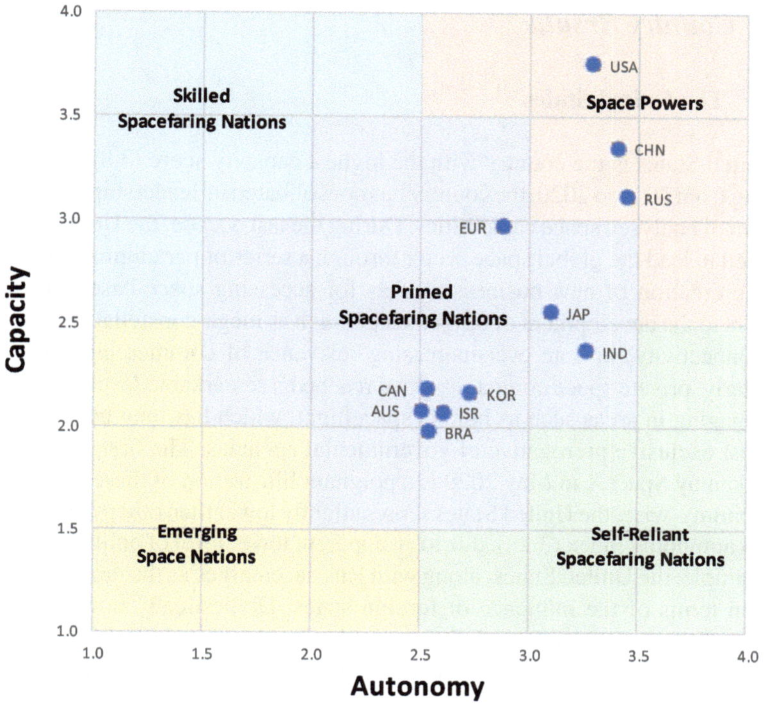

Fig. 4.25 Positioning of actors in the spacepower matrix with taxonomy

completely different league. They feature a unique combination of high levels of capacity and high levels of autonomy that no other country can currently match. Within this first group, the United States emerges as slightly more capable but also slightly less autonomous than China and Russia. The second cluster group includes Europe, Japan and India. While Europe emerges as considerably more capable than the other two, it also presents much lower levels of autonomy. The third and final cluster that emerges from our analysis includes Australia, Brazil, Canada, Israel and South Korea. All these countries present considerable room for improvement in terms of both capacity and autonomy, especially in comparison to the performance of a country such as Japan. In the following pages, we present a short country-by-country analysis of our results.

4.4.1 Country Results

4.4.1.1 The United States

The United States is the country with the highest capacity score (3.8). In the period spanning from 2011 to 2020, the country has consolidated its leadership across the full spectrum of analysed space capabilities. During the last decade, the United States has continued to lead the global space sector through a series of paradigm shifts (ranging from the creation of new business models for accessing space based on reusable launchers to the development of new concepts such as mega-constellations for broadband connectivity) and an ever-increasing relevance of commercial space actors. Particularly, private space companies have reached a remarkable level of sophistication, engaging in areas such as human spaceflight, which has long been considered an almost exclusive prerogative of governmental agencies. The first crewed launch undertaken by SpaceX in May 2020 is a poignant illustration of these new frontiers.

Autonomy-wise, the United States scores slightly lower than other space powers in the total autonomy index (3.28), due to an apparent lower level of political autonomy. In our sample, the United States, along with Russia, emerges as the most autonomous nation in terms of the influence of foreign states. Domestically however, it ranks the lowest in terms of military influence and second to last in terms of autonomy from business corporations, both in external and internal decisions (see figures 4.17 and 4.19). This reflects the strong influence exercised by the military-industrial complex on a variety of issues in the space-related decision-making processes. Military involvement in US space activities has been apparent since the onset of the US space journey and is also evident from the near hegemony of military-oriented works in the American space literature (see Ch. 2.3.1). In today's exceedingly challenging security environment, US dependence on space-based capabilities has significantly increased, contributing to making the US defence establishment particularly influential at every level of the decision-making processes. US efforts to maintain its dominance in space and ensure domain stability to its favour have a huge impact on the making of the national policy agenda. Private corporations, in turn, have a significant influence on shaping national space policies, regulations, and the basic parameters of the space programme. In our sample, the United States (along with Europe) ranks the least, scoring 2.2 in terms of autonomy from corporations in internal decisions. Once more, this aligns with the extensive body of literature on lobbying and regulatory capture in the United States, as explored in Chapter 2.2.2. This comes as no surprise, since the aerospace industry is among the most influential lobbying groups in Washington. This observation is in line with our earlier characterisation of state-business relations in the U.S. as a particular form of "competitive symbiosis" where business elites tendentially prevail. Our analysis of space policy further corroborates this perspective, underscoring that power and stateness are not inherently synonymous concepts. A country may possess a powerful government, even in the presence of a relatively low level of stateness.

4.4.1.2 China

Risen as the second most-capable space power (3.36) of the twenty-first century, China has been accumulating a series of incredible feats and achievements in the space domain for the last two decades. During the years from 2011 to 2020, China has commissioned multiple launch vehicles, orbited two human-rated space laboratories, completed a state-of-the-art GNSS (i.e., the BeiDou navigation satellite system), landed robotic spacecraft on both the Moon and Mars, while building an impressive satellite fleet for both governmental and military applications, and ended the decade close by starting the deployment of a new modular space station (Tiangong) comparable in size to the Soviet "Mir." China's space accomplishments also prove key contributors to its broader socio-economic policies, security and defence strategies and diplomatic initiatives. Additionally, within the domestic context, they serve as an important validation of China's political system and ruling party, while also serving as a tangible representation of the highly anticipated realisation of the Chinese dream of achieving a great national rejuvenation (Aliberti, 2015; Pollpeter et al., 2015).

Together with its high level of both hard and soft capacity, what is also remarkable is that China demonstrated to be slightly more autonomous than the United States in the total autonomy index (3.4). Undoubtedly the Chinese party-state enjoys particular freedom of manoeuvre in formulating its domestic and international policies, which will enable the country's leadership to pursue even higher space ambitions in the upcoming decade. However, China's space policymaking process proves to be beset by bureaucratic divisions and issues pertaining civil-military relations, which are a typical feature of Chinese "authoritarian" politics. Indeed, already in the late 1980s China was being described as a prime example of "fragmented authoritarianism" where, behind a monolithic facade, power was divided among various government agencies, state-owned enterprises, and other groups with competing interests, resources, and influence (Lieberthal and Oksenberg, 1988). Recent studies indicate that this characteristic may have persisted as China embraced a state-capitalist model over the past two decades; in fact, it may have even increased. Conflicts and negotiations do occur in the policy-making process, and the party-state elites may not be as autonomous as their official image suggests (Brødsgaard 2016). Even in the realm of space policy experts find that "fragmented authoritarianism continues to be the most robust framework available" for understanding policy-making in China (Besha, 2010). In sum, the military establishment and top management of state-owned enterprises play a significant role as agents of influence in space-related decisions of China as depicted in Figure 4.17 and 4.19.

4.4.1.3 Russia

With space capabilities inherited from the Soviet Union, during the period from 2011 to 2020, Russia has somewhat experienced a decline in some of the major aspects of its spacepower potential. As for the capacity dimension, Russia (3.13) has been surpassed by China. Specifically, Russia has lost ground on activities such as space science and exploration, with no single Russian mission leaving Earth's orbit for a decade. This situation has been the inevitable by-product of the persistence of deep-rooted problems (such as an out-dated industrial base, gaps in quality control, excessive manufacturing capacity, low labour productivity, recurrent administrative reforms etc.) combined with a series of negative external factors outside the direct control of Russia's space authorities (in particular a tense political situation in the international arena and the still-ongoing financial and economic crisis). The combination of these factors has caused a number of important consequences for the Russian space programme, the most prominent of which are the delay/cancellation of several programmes (e.g., Venera-D, Venera-Glob, and Phobos-Grunt for planetary exploration); a continuous string of launch and spacecraft failures occurred between 2010 and 2020); a substantial reduction in the space budget; as well as a reduced presence on the commercial markets.

In spite of these remarkable drawbacks, Russia has nevertheless proved strong on the autonomy side of spacepower: it emerged as the most autonomous country (3.4) among the 11 under review. Like China, Russia has been conducting multiple off-the-edge actions with in-orbit manoeuvres free from consequences, while showcasing its leadership and independent decision-making in international space diplomatic fora. Once again, it is worth noting that Russia, like China and the US, scores the highest in the hard autonomy subdimension and in political autonomy from foreign nations. Like the other space powers, Russia is self-sufficient in space. However, similar to China and the US, the situation changes when it comes to the autonomy of the state from the military and business corporations. As a matter of fact, Russia exhibits its own version of "fragmented authoritarianism." As mentioned in Chapter 2.2.2, even in the Putin era the "party of power" and the government are subjected to intensive lobbying. Simultaneously, the Russian state is structurally dependent on private capital and capital holders. Overall, a number of groups are involved in the policy-making process, including the government bureaucracy, state-owned enterprise leadership, the military, and the so-called oligarchs. This relative fragmentation of power may significantly vary across sectors and policy areas, and in the context of space policy, may be less overt. This aligns with our characterisation of Russia's "competitive symbiosis" as one where political elites tendentially prevail over divergent forces and interests. However, evidence suggests that competing interests are indeed vying to influence the policy formation process in the space policy domain. In our 11-country sample, Russia ranks third to last in terms of state autonomy from the military and fourth to last in terms of autonomy from corporations (see Figure 4.19).

4.4.1.4 Europe

Despite obtaining a capacity score (3.0) slightly lower than China, Russia, and the United States, Europe has demonstrated state-of-the-art capabilities across a wide spectrum of activities. The continent proves one of the most experienced actors in the international arena and operates a highly regarded space programme. Through its different constituents—the EU, ESA, EUMETSAT and their member states—it has mastered a wide array of capabilities and achieved many successes with breakthrough missions and programmes, such as the ESA-led Rosetta or BepiColombo missions, the EU's Copernicus and Galileo flagship programmes, and world-leading commercial solutions for telecommunications and launch services (Aliberti et al., 2020). That said, Europe still lags behind prominent "space powers" in several critical fields such as SSA or space for military and security applications, with several soft capacity scores negatively impacting its overall spacepower performance. Similarly, it does not, as yet, possess technological autonomy across the entire spectrum of space capabilities (e.g., crewed operations). Overall, Europe's score in the technical autonomy sub-dimension (3.4) is balanced out by a meagre 2.3 on the political autonomy sub-dimension, resulting in a total autonomy value of 2.9. While relatively more autonomous from the military than the US, China and Russia, Europe lacks full political autonomy vis-à-vis domestic corporations and is rather vulnerable to the inducements of foreign actors when it comes to both external and internal decisions. The dependencies embedded in the current political setting prove to be a specific feature of the European polity. Because of the complicated tangle of relationships, institutions, and multi-level sharing of competences, Europe as a whole indeed lacks the fundamental features defining autonomy over space matters. Specifically, the interplay between national, intergovernmental and communitarian frameworks has created an institutional misalignment, making decision-making processes inherently cumbersome. The problem is, however, more fundamental than the existence of "parallel" authority structures at European level. In fact, even when European institutions are endowed with formal authority, the de facto power to define pan-European policies has ultimately remained in the hands of the various national entities. As a result, for the definition of their space programmes and policies, both the EU and ESA are ultimately dependent on a "green light" from their member states—an aspect that can, in turn, hamper the development of major European initiatives and ultimately impair Europe's ability to develop a clear and coherent strategy for the future (Aliberti, Cappella and Hrozensky, 2019). Generally speaking, values for the 94 indicators suggest that Europe is relatively more solid with respect to the "hard" components of spacepower.

4.4.1.5 Japan

Japan also features as a major space actor, not so distant from Europe. Japan has been the undiscussed space leader in Asia for almost five decades and during the last one, it has continued building sophisticated space capabilities, especially in strategic macro-areas such as "satellite applications" and "enabling and support." Notably, Japan's soft capacity score roughly equals its hard capacity score, with highest values associated with "environment and resources" indicators (e.g., integration of space technology in terrestrial activities including agriculture and meteorology): however, Japan's overall capacity score (2.57) leaves room for strategic improvement in the years to come and urges for an enhancement of the hard capacity components of spacepower. While the pursuit of end-to-end capabilities and the development of more complex programmatic initiatives in various macro-areas (e.g., safety and security) is already under way, it is undeniable that the relative size of Japan's space programme will increasingly matter, especially when comparing Japan's space efforts to the ones led by continental-size actors such as US, China, and even Europe (Aliberti & Hadley, 2020). As these countries continue to increase the resources dedicated to space activities along with their level of ambition, Japan will find it hard to keep pace with their developments and retain its competitive edge in all domains of activities (and secure a good position in the spacepower matrix). Indeed, notwithstanding the remarkable synergies the country is creating among the various facets of its programme, as long as the dedicated resource remains within the same order of magnitude, it will be hard for Japan to effectively fulfil all its ambitions. This implies that the country will be inevitably confronted with both the need to prioritise different objectives and strike a proper balance between multiplying the benefits of enhanced cooperation and the level of independence it wants or can achieve.

With regard to the more specific results in the autonomy dimension, Japan obtained an above-average autonomy score of 3.14 in the total autonomy index, performing slightly better on the "technical autonomy" sub-dimension than on the "political autonomy" one. Specifically, responses to our survey suggests that Japan lacks full political autonomy when it comes to formulating policies and selecting partners within the space policy domain (mostly, though not only, vis-à-vis foreign influences—supposedly those coming from its most dependable partner, the United States). In terms of autonomy from business corporations, Japan falls in the median range within our sample. In this regard it is worth recalling that, as mentioned in Chapter 2.2.2, studies on state-business relations in Japan have highlighted that the extent of state capture by business corporations varies significantly depending on the policy area or sector of state activity under consideration. Based on our data, it can be observed that in the field of space policy, the Japanese state appears to exhibit higher autonomy from business interests compared to other areas. Looking into the future, it can be expected that Japan's international space cooperation will continue to deepen, particularly with the US, but also with other like-minded partners (e.g., European countries and institutions as well as Australia and India) that may provide Japan with alternative solutions and help it addressing the perennial concerns over both entrapment and abandonment from the US (Snyder, 2007).

4.4.1.6 India

India's overall capacity score (2.39) is consistent with the evolutionary path of a fast-growing, though still-developing, space actor: its major weaknesses lie in the hard capacity sub-dimension. Specifically, India will need to enhance its space capabilities in a number of critical macro-areas, including "space safety and security" (encompassing the acquisition of SSA and counter-space technologies), where it gained a poor 1.33 out of 4 as well as "science and exploration," particularly human spaceflight activities, where it obtained a very modest 1.4. At the time being, its soft capacity score (2.66) exceeds its hard capacity score (2.09) by almost 0.6 points. The development of downstream applications and integration of space assets in national policies and infrastructure (particularly in the socio-economic policies" macro-area) has indeed been one of the most developed features of the Indian space activities since their inception in the 1960s.

That said, it is noteworthy that India has been assigned the highest score (3.50) for political autonomy in the empirical assessment presented in this study, signifying that it is one of the most autonomous countries in terms of both internal and external decision-making. This invites a further set of observations. Similar to other countries sporting lower levels of capacity, India is endowed with comparatively higher levels of autonomy; it is also more politically autonomous than it is technically autonomous, or self-sufficient. Political autonomy, in other words, seems to compensate for deficiencies in other areas. This aligns with studies discussed in Chapter 2.2.1, particularly Peter Evans' characterisation of India as an intermediate case along a continuum of state-society relations ranging from *immersion* to *insulation*: in the context of space policy, the balance tilts towards insulation. Furthermore, as we have categorised India as a case of "mutualistic symbiosis" between business and the state, it is important to note that, on balance, mutualism occurs today under stronger state control, particularly during the Modi government. Although space policy governance has encountered occasional bouts of cronyism in the past, presently it displays a noteworthy level of insulation from undue influences. This positive trend bodes well for India's potential as a spacepower, positioning the nation as a Primed Space Actor on its path towards becoming a fully-fledged space power in the future.

4.4.1.7 South Korea

South Korea's overall positioning within the matrix is consistent with the ambitions of a still emerging but fast-developing space actor. South Korea obtained a score of just 1.52 for the hard capacity index and of 2.39 points for the technical autonomy one though excelling in few strategic macro-areas such as space infrastructure. South Korea has been severely penalised by a low hard capacity score (1.57), particularly in the macro-areas "space safety and security," "science and exploration," and "enabling and support." Whereas the Korean Aerospace Research Institute (KARI), in charge of the civilian space programme, started development activities for its Nuri launcher (also known as KSLV-II) in 2010, no successful orbital launch was undertaken during the considered timeframe, also due to the several strains in the foreseen cooperation with Russia and the United States (Aliberti et al., 2021).

In spite of the currently low scores, South Korea overall stands out as a country with strong potential for growth thanks to an ever-growing political commitment to the expansion of national space activities and a solid tech-industrial base. Korea's growing space ambitions are evidenced by governmental policies as "KARI Future Vision 2050" and the "Basic Plan for Promotion of Space Development" (Aliberti et al., 2021). Driven by the fourfold overarching goal of inspiring people, fostering industry and employment, advancing strategic areas, and contributing to the national interest, this latter document spells out important goals, including the launch and further upgrading of domestic launch capabilities (KSLV-II); the development of a public multi-purpose GEO communication satellite by 2027; the lunar landing of a national spacecraft launched on a domestic vehicle by 2030; as well as the establishment of the Korea Positioning System (KPS) by 2035.

Beyond the capacity dimension, South Korea proved rather resilient to domestic and international pressures, as demonstrated by an overall political autonomy score of 3.13. To be better appreciated, this observation must be put into context. The largest of the so-called "Asian Tigers," South Korea is known for its success as a developmental state. As mentioned in Chapter 2.2.1, the country represents the epitome of the "embedded autonomy" concept, where a well-structured and coherent state is closely interconnected with the domestic business community through institutionalised channels for continuous negotiation of goals and policies. This helps qualifying the substantial level of autonomy exhibited by the South Korean state in the realm of space policy. In fact, although South Korea's first KSLV-I rocket "Naro" was developed and launched with Russian assistance in 2013, the Koreans have worked hard since to enhance their technical autonomy (self-sufficiency) until they independently developed and launched the KSLV-II rocket "Nuri" and the KPLO lunar orbiter "Danuri" in the Summer of 2022. These successes further solidified the demands long made on the government by South Korean space industry leaders, who called for the establishment of a national aerospace agency that could help streamline decision-making and enhance coordination among various stakeholders, including government agencies, research institutions, and private companies. The private industry's interest in the matter aligned with the government's traditional concerns regarding North Korea's missile technology disguised as satellite launches. In November 2022, the country's newly elected President Yoon Suk-yeol – who has an assertive stance towards North Korea and had pledged to expand space programmes during his election campaign – announced plans to establish the Korea Aerospace Agency (KAA) by the latter half of 2023. Overall, the high level of embeddedness, or "mutualistic symbiosis" between the South Korean state and the space industry has led to the presently planned activities and will certainly help the country to substantially advance both its hard and soft capacities, eventually catching up with Japan and India in the journey from its present status of Primed Space Actor to that of Space Power.

4.4.1.8 Canada

Canada has been attributed under-average scores for both the autonomy and the capacity dimensions (2.63, 2.26), despite gaining a perfect 4/4 with respect to the "infrastructure" area included in technical autonomy sub-dimension and collecting some valuable performances concerning political autonomy, particularly from the influence exercised by the military-industrial complex. In addition, regular consultations with government agencies, including the Canadian Space Agency (CSA), along with extensive public-private partnerships within the space sector, create ample opportunities for the Canadian space industry, in particular defence contractors, to actively engage in lobbying activities to promote their interests and priorities. However, given its specialisation in niche space products such as Canadarm contribution to the ISS programme and space-based radar systems, Canada has been mostly relying on other actors' space capabilities: unexpectedly, Canada's dependence on external agents has produced a significant impairment in the overall autonomy dimension. Indeed, Canada obtained a political autonomy score of 2.94 and proved particularly vulnerable to foreign nations' invitations and interventions.

Since it became one of the world's first space nations in 1962 with the launch of its own Alouette-1 experimental satellite into orbit, Canada has maintained close bilateral cooperation with the United States in the fields of space science and technology, security, and defence. This naturally makes Canadian policy-making processes particularly sensitive to US preferences, which, together with ESA, remains Canada's most dependable partner in space. The significant dependence on cooperation with the United States represents both an asset and a possible hindrance in the development of its future national space strategies.

4.4.1.9 Israel

With a total capacity index of 2.1 and autonomy index of 2.6, Israel emerges from the assessment as a primed spacefaring nation. Like South Korea, Israel has failed to develop state-of-the-art capabilities in several macro-areas including "space exploration" and "space safety and security," receiving a total hard capacity score of 2.17.

This is mainly because Israeli space programme, established in the early 1980s to serve domestic institutional needs in the area of telecommunications and remote sensing, is still largely focused on developing and launching indigenous reconnaissance and communication aboard national transportation systems. In this specific area, over the past decade Israel undertook few successful launches from the Palmachim launch base while also working towards adapting the payload sizes of its satellite system to fit its small Shavit launcher through miniaturisation and increased development of electric propulsion technology. The focus of the Israeli space programme has expanded in more recent years, with first forays made into the area of exploration through the Beresheet lunar lander project.

With regard to the autonomy dimension, Israel obtained a score of 2.4 for hard (technical) autonomy, despite performing slightly better on the political autonomy side (2.9). While national space policy is defined by the civilian Israeli Space Agency (ISA), the military-industrial complex has an important influence through its leading role in development. Defence contractors such as Israel Aerospace Industries (IAI) and Rafael Advanced Defense Systems, are actively involved in developing and manufacturing space-related technologies, satellites, and launch systems. These companies' interests are closely intertwined with national security considerations, encompassing the military's needs for space-based assets and technologies. This symbiotic relationship between private corporations and national security plays a pivotal role in shaping Israel's space policy decisions.

4.4.1.10 Australia

With regard to Australia's position within the spacepower matrix (2.85, 2.35) it is worth noting that Australia performed much better with respect to soft (political) dimensions than to hard(technical) dimensions of spacepower. Particularly, Australia ranked second in the political autonomy special ranking, gaining a score of 3.4. Australia showed great resilience to exogenous pressures, especially from domestic corporations. Also, Australia crossed the 2.5-point threshold for soft capacity (2.8), with the highest scores being associated with areas including "security" and "environment and resources." Overall, the country has been penalised by relatively low scores for hard autonomy (2.2) and hard capacity (1.4).

Despite its early involvement in space activities with the launch of WRESAT in 1967 from the Woomera Test Range, the deployment of Australis-OSCAR 5 in 1970 as well as its important contribution to the Apollo 11 mission with the Parkes Radio Telescope Observatory, since the early 1980s Australia's space journey incurred a period of relative stagnation and never really took off. Several space development efforts were cancelled due to limited financing and an uncertain strategic direction. As a result, for most of its space history Australia relied on foreign nations' assets, particular the US ones, to develop its soft capacities. In recent years, however, the country's activities have respawned in a new direction, now oriented towards the industry as well as new institutional programmes (both civil and military): from the establishment of new governmental entities such as the Australian Space Agency (ASA) and the support of private companies, to the reform of legislation and policies, and the development of new defence activities (Aliberti et al., 2021). A new focus is put on commercial industry as a driver for jobs and economic growth. In this respect, the Australian civil space strategy 2019–2028 recognises the importance of space industry in diversifying the economy, developing national capability, and inspiring and improving lives of all Australians. The declared objective is to triple the size of the space sector from $3.9 billion to $12 billion and grow the segment from around 10,000 jobs to 30,000 jobs by 2030, with further job creation and economy growth expected from spillover effects (ASA, 2019). The very creation of ASA responds to a specific belief of enabling industry to deliver innovative solutions, rather than managing institutional space activities. Developing proper in-house space capabilities, be them industry- or government-led, can be expected to push Australia towards the upper-right quadrant of the matrix, at a close distance from prominent space actors like Japan, India and Europe.

4.4.1.11 Brazil

With a capacity index of 2.0 and an autonomy index of 2.5, Brazil appears to have been particularly penalised by the capacity-autonomy coupling. Despite dating back to the 1960s, Brazilian space activities have never fully taken off, historically pursued with limited budgetary commitments in the context of often precarious socio-economic conditions. Within the hard capacity sub-dimension, it is worth highlighting that the country implemented a launcher programme in the past but was unable to undertake a successful launch during the past decade. Also, Brazil initiated spare satellite programmes, though no major capacity (both hard and soft) has been built up in the years from 2011 to 2020. Brazil's above-average score of 3.1 for the political autonomy sub-dimension partly compensates for the poor performance on hard (technical) autonomy (1.9), with notable weaknesses lying in areas including launch and crewed operations. Notably, Brazil emerged as the least capable country of the assessment with respect to the hard capacity sub-dimension (1.25). Even today, the Brazilian space programme continues to be driven by local needs with a focus on satellite applications, particularly remote sensing for monitoring the large national territory and satellite communication for connecting rural areas. This niche focus explains the poor performance on other hard capacity indicators, such as "science and exploration" and "safety and security."

4.4.2 Comparative Results

All things considered, when we conflate our data into a single composite index, only three actors emerge as full-fledged space powers, in the order: the United States, China and Russia—with Europe following at considerable distance.

With significantly high autonomy and capacity, the United States (3.40, 3.76) emerges as the indisputable master of all things space, getting closer to the space power ideal. The data confirm the decline of Russia (when compared to the old standard of the Soviet Union) and the parallel rise of China, which over the past decade has been accumulating a series of incredible feats and achievements that will enable it to pursue even higher space ambitions in the current decade.

The competition between China and the US emerges as the dominant scenario. China, fulfilling its ambition of "national rejuvenation," will inexorably continue to pursue and expand its interests in space, marching steadily and strategically towards catching-up with what the US has achieved. According to most commentators (Bowe, 2019; Moltz, 2012; Pollpeter et al., 2015; Seedhouse, 2023; Stokes et al., 2020; Zhang, 2013), irrespective of China's declared interest in avoiding a space race or a strategic arms race, the mere expansion of its space ambitions is bound to fuel a self-propelling mechanism that will eventually lead to confrontation. It is interesting to note, in this respect, that China enjoys higher levels of political autonomy than the US, a dimension where the US scores the lowest, due in particular to the influence exercised by the military and domestic corporations on the decision-making process.

It may appear tempting (though not conclusive) to relate this to the difference in political regimes, with an autocratic, insulated regime quite successfully challenging a democratic, open one. This, however, would be an oversimplification. In fact, the distinction seems to confirm that states can pursue various strategies and paths in their quest for the status of a space power. While the attainment of high levels of autonomy and capacity is undeniably crucial for gaining entry and maintaining membership within the "club," this goal can be achieved through varying combinations of hard and soft factors, where relative shortcomings in one aspect are compensated for by relative surpluses in others.

Our findings also offer some food for thought when it comes to comparing the US, which is still the indisputable leader in the space power circle, and Europe, which lags a distant fourth. Although the United States scores rather badly in political autonomy, it owes its persistent leading position to its enormous strength in all the other sub-dimensions. Europe on the other hand, owes its comparatively lower position to its negative scores in both our soft dimensions: to catch up with the US and other full-fledged space powers Europe needs improve its standing in both soft capacity and political autonomy.

As for the other countries included in our sample, those to watch more closely for their capacity and autonomy performances are Japan (3.14, 2.17), and India (2.4 3.25), which are both in the process of increasing their space capacities (hard and soft) while maintaining the relatively good scores in the dimension of autonomy, especially technical autonomy.

Interestingly, none of the countries in our sample fits the quadrant named "skilled spacefaring nations" (highly capable space nations with rather limited autonomy). On the contrary, most states appear to be more autonomous (especially in the hard autonomy realm) than capable—thus falling in the "self-reliant space-faring nations" quadrant. Among the space actors falling in this quadrant a general trend can be observed: lower hard capacity scores tend to go together with lower technical autonomy scores. The strong correlation between hard capacity and technical autonomy can be explained by the "coupling" of the two sub-dimensions. As the world transitions towards more commercially driven space activities, more and more countries will be able to acquire capacity by simply purchasing hardware and services, thus enabling a decoupling of capacity and autonomy features. Until then, countries which do not invest in developing endogenous capacity are less likely to be autonomous, and vice versa.

References

Al-Ekabi, C. (2016). *The Future of European Commercial Spacecraft Manufacturing* (ESPI Report 58). European Space Policy Institute.

Aliberti, M. (2015). *When China Goes to the Moon...* Springer.

Aliberti, M. (2018). *India in Space—Between Utility and Geopolitics.* Springer.

Aliberti, M., & Hadley, S. (2020). *Securing Japan. An assessment of Japan's strategy for Space* (ESPI Report 74). European Space Policy Institute.

Aliberti, M., Cappella, M., & Hrozensky, T. (2019). *Measuring Space Power: A Theoretical and Empirical Investigation on Europe*. Springer.

Aliberti, M., Hrozensky, T., & Bataille, M. (2020). *European Space Strategy in a Global Context* (*ESPI Report* 75). European Space Policy Institute.

Aliberti, M., Renčelj, M., Poirier, C., Hallet, L., & Hermes, M. (2021). *Emerging Spacefaring Nations. Review of Selected Countries and Considerations for Europe* (ESPI Report 79). European Space Policy Institute.

Aoki, S. (2009). Current Status and Recent Developments in Japan's National Space Law and its Relevance to Pacific Rim Space Law and Activities. *Journal of Space Law, 35,* 363–438.

ASA. (2019). *Advancing Space, Australian Civil Space Strategy 2019–2028*. Australian Space Agency. https://bit.ly/3xlvVPy (Accessed on 21 May 2022).

Besha, P. (2010). Policy making in China's space program: A history and analysis of the Chang'e lunar orbiter project. Space Policy. 26(4) 214-221.

Bowe, A. (2019). *China's Pursuit of Space Power Status and Implications for the United States*. U.S.-China Economic and Security Review Commission.

Brødsgaard, K.E. (Ed.). (2016). Chinese Politics as Fragmented Authoritarianism: Earthquakes, Energy and Environment. Routledge.

Casarini, N. (2009). *The Remaking of Global Order. The Evolution of Europe-China Relations and its Implications for East Asia and the United States*. Oxford University Press.

Chatterjee Miller, M. (2013). India's Feeble Foreign Policy. *Foreign Affairs, 92*(3), 14–19.

Clark, S. (2017). *Soyuz Rocket Lifts Off with 73 Satellites*. Spaceflight Now. https://spaceflightnow.com/2017/07/14/soyuz-rideshare-launch/ (Accessed on 30 May 2022).

CSIS. (2019). *Spaceports of the World*. Aerospace Security Project. Center for Strategic and International Studies. https://aerospace.csis.org/wp-content/uploads/2019/03/190313__SpaceportsOfTheWorld.pdf (Accessed on 10 May 2022).

D-Orbit. (2021). *Satellite Launch and Deployment*. D-Orbit. https://www.dorbit.space (Accessed on 30 May 2021).

ESA. (2020). *SSMS Inaugural Flight on Vega*. European Space Agency. http://www.esa.int/ESA_Multimedia/Videos/2020/06/SSMS_inaugural_flight_on_Vega (Accessed on 10 June 2021).

ESA. (2021a). *European Service Module*. European Space Agency. http://www.esa.int/Science_Exploration/Human_and_Robotic_Exploration/Orion/European_Service_Module (Accessed on 30 May 2021).

ESA. (2021b). *Rosetta*. European Space Agency. https://www.esa.int/Enabling_Support/Operations/Rosetta (Accessed on 30 May 2021).

ESPI. (2020). *Artemis Accords: What Implications for Europe?* (ESPI Executive Brief 46). European Space Policy Institute.

European Commission. (2022). *An EU Approach for Space Traffic Management. For a Safe, Secure and Sustainable Use of Space*. Defence Industry and Space—European Commission. https://defence-industry-space.ec.europa.eu/eu-space-policy/eu-space-programme/eu-approach-space-traffic-management_en (Accessed on 10 August 2022).

EUSPA. (2021). *EU Space on a New Trajectory with Launch of EUSPA*. European Union Agency for the Space Programme. https://www.euspa.europa.eu/newsroom/news/eu-space-new-trajectory-launch-euspa (Accessed on 10 August 2022).

Fournier, M., et al. (2018). Past, Present, and Future of the Satellite-Based Automatic Identification System: Areas of Applications (2004–2016). *WMU Journal of Maritime Affairs, 17*(3), 311–345.

Foust, J. (2021). *Europe Considering Concepts for Human Spaceflight*. SpaceNews. Available at: https://spacenews.com/europe-considering-concepts-for-human-spaceflight/ (Accessed on 10 August 2022).

Foust, J. (2022). *U.S. Looking to Encourage More Countries to Join ASAT Testing Ban*. Space News. https://spacenews.com/u-s-looking-to-encourage-more-countries-to-join-asat-testing-ban/ (Accessed on 29 October 2022).

IDA. (2020). *Measuring the Space Economy. Estimating the Value of Economic Activities in and for Space*. Institute for Defense Analyses.

ISRO. (2021). *PSLV-C37 Successfully Launches 104 Satellites in a Single Flight*. Indian Space Research Organisation. https://www.isro.gov.in/pslv-c37-successfully-launches-104-satellites-single-flight (Accessed on 30 May 2021).

Hayward, K. (2011). *The Structure and Dynamics of the European Space Industry Base* (ESPI Perspectives 55). European Space Policy Institute.

JAXA. (2021). *JAXA Hayabusa2 Project*. Japan Aerospace Exploration Agency. https://www.hayabusa2.jaxa.jp/en/ (Accessed on 30 May 2021).

Johnson-Freese, J., & Erickson, A. S. (2006). The Emerging China-EU Space Partnership: A Geo-technological Balancer. *Space Policy.*, *22*(1), 12–22.

Jones, A. (2021). *Astronauts Complete First Chinese Space Station Spacewalk*. SpaceNews. https://spacenews.com/astronauts-complete-first-chinese-space-station-spacewalk/ (Accessed on 10 August 2022).

Kallender, P. (2016). *Japan's New Dual-Use Space Policy. The Long Road to the 21st Century. Asie Visions, 88*(1).

Krebs, G. D. (2021). *Kosmos 2519 / Kosmos 2521 / Kosmos 2523*. Gunter's Space Page. https://space.skyrocket.de/doc_sdat/kosmos-2519.htm (Accessed on 30 May 2021).

LINZ. (2022). *Satellite-Based Augmentation System*. Land Information New Zealand. https://www.linz.govt.nz/data/geodetic-services/satellite-based-augmentation-system (Accessed on 30 May 2022)

Logsdon, J. M. (2021). *Space Commercialization during the Reagan Administration*. NASA History Symposium: NASA and the Rise of Commercial Space. Virtual Symposium held on 13 April 12021.

Martin, M., Pfrang, K., & Weeden, B. (2021). *Chinese Military and Intelligence Rendezvous and Proximity Operations*. Secure World Foundation. https://swfound.org/media/207179/swf_chinese_rpo_fact_sheet_apr2021.pdf (Accessed on 30 May 2022).

Mazzucato, M., Douglas, Robison, K. R. (2016). *Lost in Space? NASA and the Changing Public-Private Eco-System in Space* (Working Paper Series). University of Sussex.

MEA. (2019). *Frequently Asked Questions on Mission Shakti, India's Anti-Satellite Missile Test Conducted on 27 March, 2019*. Ministry of External Affairs, Government of India. https://www.mea.gov.in/press-releases.htm?dtl/31179/Frequently_Asked_Questions_on_Mission_Shakti_Indias_AntiSatellite_Missile_test_conducted_on_27_March_2019 (Accessed on 11 July 2022).

Moltz, J. C. (2012). *Asia´s Space Race. National Motivations, Regional Rivalries, and International Risks*. Columbia University Press.

NASA. (2021a). *Phobos-Grunt*. National Aeronautics and Space Administration. https://solarsystem.nasa.gov/missions/phobos-grunt/in-depth (Accessed on 11 July 2021).

NASA. (2021b). *On-Orbit Servicing, Assembly, and Manufacturing (OSAM)*. National Aeronautics and Space Administration. https://nexis.gsfc.nasa.gov/osam (Accessed on 11 July 2021).

NASA. (2021c). *Landsat Homepage: Landsat Science*. National Aeronautics and Space Administration. https://landsat.gsfc.nasa.gov/ (Accessed on 13 July 2021).

Newberry, R. D. (2003). Latin American Countries with Space Programs: Colleagues or Competitors? *Air & Space Power Journal, 17*(3), 39–48.

O'Callaghan, J. (2020). *Europe's First Mars Rover Delayed by Two Years*. Nature. March 12, 2020.

OECD. (2021). *Public-Private Relations in the Space Sector Lessons Learned for the Post-Covid-19 Era*. OECD Science, Technology and Industry Policy Papers 114. Organisation for Economic Co-operation and Development.

Park, S. (2021). *South Korean Leader Vows 'Landing on the Moon by 2030.'* SpaceNews, https://spacenews.com/south-korean-leader-vows-landing-on-the-moon-by-2030/ (Accessed on 20 July 2022).

Peoples, C. (2013). A Normal Space Power? Understanding 'Security' in Japan's Space Policy Discourse. *Space Policy, 29*, 135–143.

Pollpeter, K., et al. (2015). *China Dream, Space Dream: China's Progress in Space Technologies and Implications for the United States*. Institute on Global Conflict and Cooperation.

Prasad, N. (2016a). Demystifying space business in India and issues for the development of a globally competitive private space industry. *Space Policy, 36*, 1–11.

Prasad, N. (2016b). Industry Participation in India's Space Programme: Current Trends & Perspectives for the Future. *Astropolitics: The International Journal of Space Politics and Policy, 14*(2–3), 234–255.

Robinson, J. (2019). *State Actor Strategies in Attracting Space Sector Partnerships: Chinese and Russian Economic and Financial Footprints*. Prague Security Studies Institute.

Seedhouse, E. (2023). *The Growing Chinese Space Threat*. SpaceNews. Available at: https://spacenews.com/the-growing-chinese-space-threat/ (Accessed on 1 March 2023).

Sheehan, M. (2007). *The International Politics of Space*. Routledge.

Silk, M. (2021). *China Is Evolving a Distinct Space Culture*. The Diplomat. https://thediplomat.com/2021/03/china-is-evolving-a-distinct-space-culture/ (Accessed on 15 July 2021).

Stokes, M., et al. (2020). *China's Space and Counterspace Capabilities and Activities*. US-China Economic and Security Review Commission.

Suzuki K. (2015). Space Security in Japan. In K. U. Schrogl, P. Hays, J. Robinson, D. Moura, C. Giannopapa (Eds.), *Handbook of Space Security*. Springer.

Swilley, S. F. (2011). *Space Power: A Theory for Sustaining US Security Through the Information Age*. School of Advanced Military Studies.

The Planetary Society. (2021a). *Chang'e-4*. The Planetary Society. Available at: https://www.planetary.org/space-missions/change-4 (Accessed on 30 May 2021).

The Planetary Society. (2021b). *Chang'e-5: China's Moon Sample Return Mission*. The Planetary Society. https://www.planetary.org/space-missions/change-5 (Accessed on 10 July 2021).

The Planetary Society. (2021c). *Beresheet2, A Private Israeli Moon Mission*. The Planetary Society. https://www.planetary.org/space-missions/beresheet (Accessed on 10 July 2021).

The Planetary Society. (2021d). *Tianwen-1 and Zhurong, China's Mars Orbiter and Rover*. The Planetary Society. https://www.planetary.org/space-missions/tianwen-1 (Accessed on 30 May 2021).

Tugnoli, M., & Wells, L. (2019). *Evolution of the Role of Space Agencies* (ESPI Report 70). European Space Policy Institute.

UCS. (2021). *Satellite Database*. Union of Concerned Scientists. https://www.ucsusa.org/resources/satellite-database (Accessed on 5 July 2021).

US Congress. (2011). *112th Congress Public Law 10*. U.S. Government Printing Office. https://www.congress.gov/112/plaws/publ10/PLAW-112publ10.htm (Accessed on 10 October 2022).

USCC. (2019). *Hearing on "China in Space: A Strategic Competition?"* US-China Economic and Security Review Commission. https://www.uscc.gov/hearings/china-space-strategic-competition (Accessed on 20 October 2022).

Van Wagenen, J. (2015). *Effective Space Solutions Secures Downpayment Funding for First Two In-Orbit Servicing Spacecraft*. Satellite Today. https://www.satellitetoday.com/innovation/2015/07/23/effective-space-solutions-secures-downpayment-funding-for-first-two-in-orbit-servicing-spacecraft/ (Accessed on 10 July 2021).

Verspieren, Q. (2020). *A Reluctant Space Safety Services Provider: The Role of the Military in Space Traffic Management*. University of Tokyo.

Wallace, W. (2005). Post-sovereign governance: The EU as a partial polity. In: Wallace, H., Wallace, W., and Pollack, M.A. (Eds.). *Policy-making in the European Union*. Oxford University Press: 483–503.

Whitford, G. (2019). *Trouble in the Stars: The Importance of US-China Bilateral Cooperation in Space*. Harvard International Review. https://hir.harvard.edu/trouble-in-the-stars-the-importance-of-us-china-bilateral-cooperation-in-space/ (Accessed on 20 July 2021).

Zeng, K. (2004). *Trade Threats, Bargaining, Retaliation, and American Coercive Diplomacy*. University of Michigan Press.

Zhang, Y. (2013). The Eagle Eyes the Dragon in Space—A Critique. *Space Policy, 29*(2), 113–120.

Chapter 5
Conclusions

This book represents an effort to understand the actual power of state actors in the space domain both in absolute and in relative terms. To achieve this objective, the book outlines a conceptual framework (Chap. 2), devises a measurement model based on the concepts developed (Chap. 3), and utilises empirical data to conduct a comparative analysis of spacepower in 11 countries between 2011 and 2020 (Chap. 4). The outcome is the spacepower matrix (Fig. 5.1), which positions and ranks all the countries studied in a two-dimensional space, utilizing capacity and autonomy as the guiding criteria. Our conceptual framework and measurement model have the flexibility to be utilised for any country and any time period. This enables a dynamic comparison of space actors at a particular point in time and facilitates the interpretation of trajectories that can lead them towards or away from the circle of space powers. While this book presents a comparative analysis of a significant number of cases, it is possible to use its methodology to create country-specific, in-depth analyses, and even identify areas that require investment to enhance the spacepower of a specific state actor. Through the cascading nature of our measurement model, it is possible to drill down to the level of our 94 individual indicators, which can highlight, for example, that a particular country's resources may be better spent improving autonomous access to satellite data and increasing the number of tech demonstrator missions, rather than enhancing the reliability of their launchers. In practical terms, this means that practitioners and policymakers now have access to a toolkit that can assist them in decision-making processes related to the development of a country's space capabilities.

Our underlying assumption is that spacepower is a form of state power. Therefore, our analysis starts with a critical evaluation of the scholarly literature on the concept of state power, and articulates it into two basic dimensions that have had divergent fortunes among analysts: (a) capacity, which is the most extensively studied dimension presently and pertains to a state's ability to enforce laws, implement policies, and achieve objectives, and (b) autonomy, which was once a prominent topic in the literature but has recently been disregarded in the analysis of both state power and

M. Aliberti et al., *Power, State and Space*, Studies in Space Policy 35,
https://doi.org/10.1007/978-3-031-32871-8_5

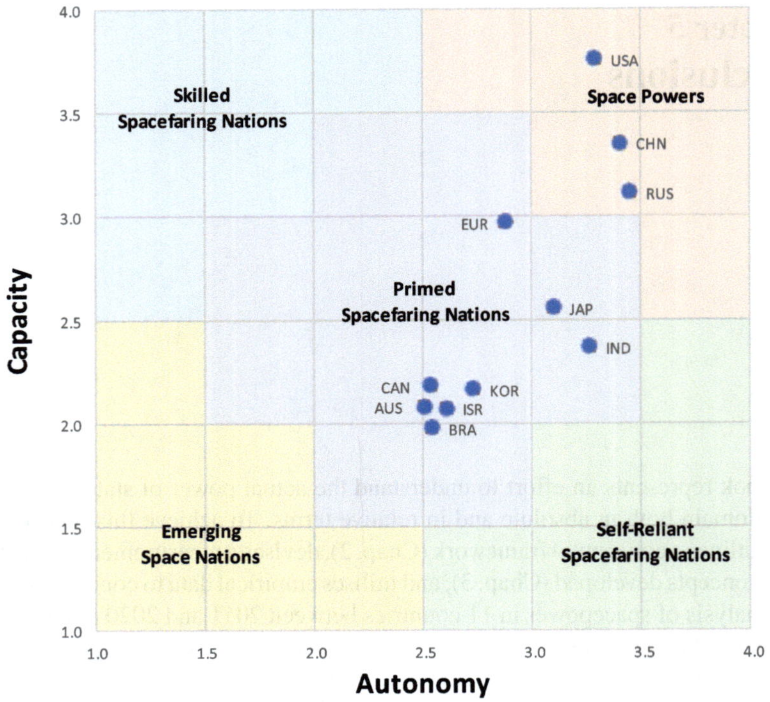

Fig. 5.1 The spacepower matrix

space power. Autonomy refers to a state's ability to formulate its policies and objectives independent of, or even against, the will of exogenous interests. In essence, the pervasiveness of the concept of state capacity has resulted in the neglect of the inherently political notion of state autonomy. Nonetheless, a political entity capable to execute decisions that it has not autonomously taken would hardly qualify as "a state" by modern standards. Similarly, a state that can engage with the space domain only if other actors, be them foreign states, the domestic military, or private corporations permit it to do so, will never truly be a space power. Consequently, we contend that spacepower should be comprehensively understood and empirically assessed as a multidimensional status pertaining to political-administrative entities that possess a high-level combination of decisional autonomy and executive capacity. We have observed that the literature on space power shares striking similarities with studies on state power. Conceptually, it is predominantly focused on capacity and pays insufficient attention to the crucial issue of political autonomy in space matters. Methodologically, it lacks a shared consensus on what to measure and how to assess the results in a comparative manner. Moreover, operationally, it has limited ability to provide useful information and guidance to policymakers.

To overcome these shortcomings we introduce a novel conceptualisation of *spacepower* in the form of a multi-dimensional matrix that considers the complex interplay between two fundamental dimensions: the ability of a state to make autonomous decisions in space-related matters, and its material capacity to execute those decisions and accomplish the associated objectives. The concept of *space power*, in turn, refers to the outcome of this measurement process and is defined as a particular status that is exclusive to those states possessing a specific level of spacepower—one in which a high degree of capacity meets a high level of autonomy.

The spacepower matrix (see Fig. 5.1) indicates the relative position of each actor at a given point in time and helps interpret their trajectories by identifying five distinct types of space actors including (1) *space powers*, (2) *emerging space nations, (3) skilled spacefaring nations*, (4) *self-reliant spacefaring nations*, and (5) *primed spacefaring nations*. To be considered a "space power," our argument goes, a nation must exhibit elevated levels of both capacity and autonomy: excelling in one of these dimensions alone is not enough. Additionally, our framework provides much-needed detail on the so-called "spacefaring nations," that is, countries that have some level of space engagement but do not possess sufficient capacity and/or autonomy to claim space power status. We argue that such countries should be classified as "skilled spacefaring nations" if they exhibit significant capacity but very low levels of autonomy. This could be the case of a country that only engages with space through the purchase of technology from foreign partners, be it other states or private corporations. Likewise, a nation should be classified as "self-reliant spacefaring nation" if, despite maintaining considerable control over space-related decisions and activities its space capacity remains limited to very few macro-areas. Although none of the countries included in our analysis fall into these two categories, recent developments, *in primis* the expansion of commercial space activities, suggest that other countries either already do or will in the future. The majority of the countries analysed in this book are classified instead as "primed spacefaring nations." In many ways, this is the most interesting type of space actor in our matrix: a primed spacefaring nation has attained "respectable" levels of both capacity and autonomy, implying that it could potentially be on its way to becoming a space power.

In order to operationalise this approach and accurately determine where an actor falls within the matrix, we developed a methodology that breaks down spacepower into its two dimensions of capacity and autonomy, and further divides them into four sub-dimensions: hard capacity, soft capacity, hard (or technical) autonomy, and soft (or political) autonomy. We also provided a detailed description of each sub-dimension's indicators and explained the measurement model behind them. Specifically, to measure the sub-dimensions of hard capacity and hard autonomy we used quantitative data from a number of specialised sources. For the sub-dimensions of soft capacity and soft autonomy, instead, we conducted surveys among an ample pool of academics and space professionals with expertise in the countries studied. The data collected from all these sources were combined into four composite indices of spacepower and presented in the form of a matrix that features a non-hierarchical classification of space actors. The heuristic and analytical utility of these indices (hard and soft capacity indices, and technical and political autonomy indices) is that they allow to collect, processes and aggregate a vast range of diverse data and offer essential tools for identifying, measuring, and comparing the strengths and weaknesses of various space actors.

From the empirical application of our theoretical and methodological framework, several interesting results emerge. More specifically, today only the United States, China and Russia meet the criteria for full-fledged space powers, as they possess high levels of both capacity and autonomy.

Europe, Japan and India emerge from the overall assessment with relatively good scores in both constituent dimensions—an achievement that, however, should not breed complacency or inactivity. As a consequence of technological development, the ever-changing requirements associated with *acquiring* or *maintaining* the status of *space power* create for these three actors an ever pressing need to further improve work on their capacity edge without overlooking their levels of autonomy, both political and technical. Space technologies change, adapt and evolve, eventually resulting in new types of space activities, new approaches to existing activities (cheaper, faster, more resilient …) and new opportunities for applications and services. These actors must hence keep pace with the rapid technological trends unfolding in the space sector if they are to avoid the risk of being left behind. Yet, our findings also suggest that these three actors might want to explore diverse approaches in terms of strategic planning and resource allocation. Europe is very close to achieving the status of a full-fledged space power, and it needs to prioritise increasing its autonomy over building up its capacity. Conversely, both India and Japan already sport sufficient levels of autonomy and should focus on improving their capacity dimension.

The remaining five countries that we assessed (Australia, Brazil, Canada, Israel, and South Korea) primarily fall into the medium–low capacity and medium autonomy range, making them relatively autonomous spacefaring nations. They face the challenge of improving substantially both their capacity and autonomy in a coordinated manner to attain the status of a space power. More specifically, this requires (a) ensuring the sustained development of hard capacities (e.g. through the upgrading of existing assets or the development of new types of activities) along with their integration into national policies and infrastructure; and (b) maintaining their levels of political autonomy while expanding technical autonomy to enable independent activity without seeking permission from external actors.

For all the considered countries, the application of our conceptual and methodological framework in the future will enable us to track changes in their position in the matrix and interpret the trajectory of these changes. By conducting comparative assessments over time, we will be able to capture the direction of change for each country, both individually and in relation to other countries. In other words, we can identify whether a nation is in the process of becoming or has the potential to become a space power, and if not, what measures need to be taken to rectify the situation. This can provide valuable insights for scholars, space professionals, and policymakers on how to evaluate and improve their relative status in the increasingly crowded and highly competitive international space arena.

Future research should expand the application of this framework to a more comprehensive range of cases. It would be especially interesting to analyse the present positioning and projected course of countries that are just beginning their foray into space activities. In this book, we deliberately chose to investigate 11 nations with considerable space involvement, which means that all of our cases are either already space powers or nations that are well on their way to becoming one (primed space-faring nations). However, our framework could also be a valuable tool in aiding the development and execution of national space strategies for emerging space nations. It would be also interesting to replicate our analysis for different time periods, to gain insight into the trajectory followed by our 11 countries to reach their current status. In particular, a more detailed examination of China's ascent into the exclusive group of space powers would likely yield very interesting results.

More broadly, the framework has been designed to enrich the ongoing discourse on space policy and strategy by providing insights into how space nations can improve their status as major strategic actors in space. This includes developing, projecting, and sustaining national capabilities that shape their role and position in the global space arena. Ultimately, the framework is meant to support evidence-based decision-making. It is therefore our hope that the information, data, and materials presented in this research will prove useful to those involved in designing policy strategies and providing policy advice to decision-makers.

Appendices

Appendix A: Major Literature on Spacepower (1988–2020)

Year	Author	Title	Definition
1988	Lupton, David E.	On Space Warfare: A Space Power Doctrine	Added
1995	Hyatt III, James L. et al	Space Power 2010	Added
1995	Mantz, Michael R.	The New Sword: A Theory of Space Combat Power	No definition
1996	Gray, Colin S.	The Influence of Space Power Upon History	Added
1997	Billman, Gregory	The Inherent Limitations of Spacepower: Fact or Fiction?	Added
1997	Newberry, Robert D.	Space Doctrine for the 21st Century	No definition
1998	Johnson, Dana J.; Pace, Scott; Gabbard, C. Bryan	Space: Emerging Options for National Power	Added
1998	Jusell, Judson J.	Space Power Theory: A Rising Star	No definition
1998	United States Air Force	Space Operations—Air Force Doctrine Document 2–2	Added
1999	DeBlois, Bruce M.	Beyond the Paths of Heaven: the Emergence of Space Thought	Added
1999	Gray, Colin S.; Sheldon, John B.	Space Power and the Revolution in Military Affairs	Added
1999	Oberg, James E.	Space Power Theory	Added
2000	Doyne, Thomas A	Space and the Theater Commander's War	No definition
2000	France, Martin E. B.	Back to the Future: Space Power Theory and A.T. Mahan	No definition

(continued)

M. Aliberti et al., *Power, State and Space*, Studies in Space Policy 35,
https://doi.org/10.1007/978-3-031-32871-8

(continued)

Year	Author	Title	Definition
2000	France, Martin E. B.	Mahan's Elements of Sea Power Applied to the Development of Space Power	No definition
2000	Hays, Peter L.; Smith, James M.; Van Tassel, Alan R.; Walsh, Guy M.	Spacepower for a new Millennium: Space and U.S. National Security	No definition
2001	Center for Security Policy	Space Power: What Is at Stake, What Will It Take	No definition
2001	Dolman, Everett C.	Astropolitik: Classical Geopolitics in the Space Age	No definition
2001	Fox, John G.	Some Principles of Space Strategy (or 'Corbett in Orbit')	Added
2001	Lambakis, Steven	On the Edge of Earth: The Future of American Space Power	Added
2001	Lambakis, Steven	Space Weapons: Refuting the Critics	No definition
2001	Logsdon, John M.	Just Say Wait To Space Power	Added
2001	Smith, M. V.	Ten Propositions Regarding Spacepower	Added
2002	Hays, Peter L.	What is Spacepower and Does it Constitute a Revolution in Military Affairs?	No definition
2003	Newberry, Robert D.	Spacepower as a Coercive Force	No definition
2003	Teets, Peter B.	Developing Space Power—Building on the Airpower Legacy	No definition
2004	Correll, Randall R.	Military Space Cooperation: Aligning the Balance of Power and Building Common Interest	No definition
2004	Moorthi, D.	What 'Space Security' Means to an Emerging Space Power	Added
2005	Correll, Randall R.; Worden, Simon P.	The Demise of US Spacepower: Not with a Bang but a Whimper	No definition
2005	Oberg, James E.	Towards a Theory of Space Power	Added
2005	Wagner, John W.	Spacepower Theory: Lessons from the Masters	No definition
2006	Fredriksson, Brian E.	Globalness: Toward a Space Power Theory	Added
2006	Gleason, Michael P.	European Union Space Initiatives: The Political Will for Increasing European Space Power	No definition
2006	Harter, Mark E.	Ten Propositions on Space Power: The Dawn of a Space Force	Added
2006	Klein, John J.	Space Warfare: Strategy, Principles, and Policy	No definition

(continued)

(continued)

Year	Author	Title	Definition
2007	Hays, Peter L.; Lutes, Charles D.	Towards a Theory of Spacepower	No definition
2007	Hertzfeld, Henry R.	Globalization, Commercial Space and Spacepower in the USA	No definition
2007	Krepon, Michael; Hitchens, Theresa; Katz-Hyman, Michael	Preserving Freedom of Action in Space: Realizing the Potential and Limits of U.S. Spacepower	Added
2007	Lutes, Charles D.; Hays, Peter L.; Smith, M. V.; Bell, Mike; Lahneman, William	Toward a Theory of Spacepower—The Influence of Spacepower on History and Implications for the Future	Added
2008	Lutes, Charles D.	Spacepower in the 21st Century	Added
2008	Noble, Michael J.	Export Controls and United States Space Power	No definition
2009	Havercroft, Jonathan; Duvall, Raymond	Critical Astropolitics: The Geopolitics of Space Control and the Transformation of State Sovereignty	No definition
2009	Peter, Nicolas	Space Power and Europe in the 21st Century	Added
2010	Farnsworth, Jeffrey A.	Space Power: A Strategic Assessment and Way Forward	No definition
2010	Paikowsky, Deganit	Membership in the Space Club: A Tool in the Hands of Medium-Sized and Small States for Empowerment and Projection of Power	No definition
2010	Peter, Nicolas	Space Power and Its Implications: the Case of Europe	Added
2011	Pfaltzgraff, Robert L.	International Relations Theory and Space Power	Added
2011	Sheldon, John B.; Gray, Colin S.	Theory Ascendant? Spacepower and the Challenge of Strategic Theory	Added
2011	Smith, M. V.	Spacepower and Warfare	No definition
2011	Swilley, Scott F.	Space Power: A Theory for Sustaining US Security Through the Information Age	No definition
2011		Toward a Theory of Spacepower: Selected Essays	Added
2012	Al-Rodhan, Nayef R. F	Meta-Geopolitics of Outer Space	Added
2012	Pozza, Maria	Emerging Space Powers: The New Space Programs of Asia, the Middle East and South America	No definition
2013	Fukushima, Yasuhito	Debates over the Military Value of Outer Space in the Past, Present and the Future: Drawing on Space Power Theory in the U.S	No definition
2013	Peoples, Columba	A Normal Space Power? Understanding 'Security' in Japan's Space Policy Discourse	No definition

(continued)

(continued)

Year	Author	Title	Definition
2013	Pfaltzgraff, Robert	International Relations Theory and Space Power	Added
2014	Anderson, Justin; Conrad, Walt; Jacobs Gamberini, Sarah	International Space Negotiations, Emerging Space Powers, and U.S. Efforts to Protect the Military Use of Space	No definition
2015	Bowen, Bleddyn E.	Spacepower and Space Warfare—The Continuation of Terran Politics by Other Means	Added
2015	Hays, Peter L.	Spacepower Theory	No definition
2015	Khan, Zulfqar; Khan, Ahmad	Chinese Capabilities as a Global Space Power	No definition
2015	Ziarnick, Brent	Developing National Power in Space—A Theoretical Model	Added
2016	Chapman, Bert	Chinese Military Space Power: U.S. Department of Defense Annual Reports	No definition
2016	Smith, M. V.	Spacepower and the Strategist	Added
2017	Czajkowski, Marek	Theory of Spacepower—A Brief Introduction	Added
2017	Paikowsky, Deganit	The Power of the Space Club	No definition
2018	Cappella, Matteo	Assessing the European Union's Spacepower	No definition
2018	Honable, Isaiah	In Search Of Space Power Theory	No definition
2018	Martindale, Michael; Deptula, David A.	Organizing Spacepower: Conditions for Creating a US Space Force	No definition
2018	Stockton, Nick	The 19th Century Argument for a 21st Century Space Force	No definition
2019	Aliberti, Marco; Cappella, Matteo; Hrozensky, Tomas	Measuring Space Power: A Theoretical and Empirical Investigation on Europe	Added
2019	Bowen, Bleddyn E.	From the Sea to Outer Space—The Command of Space as the Foundation of Spacepower Theory	No definition
2019	Klein, John J.	Understanding Space Strategy—The Art of War in Space	No definition
2019	Moltz, James Clay	The Changing Dynamics of Twenty-First-Century Space Power	No definition
2019	Satyanath, Pranav R.	Space Power and Space Warfare: A Review	No definition
2019	Townsend, Brad	Space Power and the Foundations of an Independent Space Force	No definition
2020	Bowen, Bleddyn E.	The Integrated Review and UK Spacepower: The Search for Strategy	Added

(continued)

(continued)

Year	Author	Title	Definition
2020	Bowen, Bleddyn E.	War in Space: Strategy, Spacepower, Geopolitics	Added
2020	Hays, Peter L.	Spacepower Theory and Organizational Structures	No definition
2020	United States Space Force	Space Capstone Publication—Spacepower: Doctrine for Space Forces	Added

Appendix B: Major Definitions of Spacepower

1988—Lupton, On Space Warfare

[…] spacepower is the ability of a nation to exploit the space environment in pursuit of national goals and purposes and includes the entire astronautical capabilities of the nation. A nation with such capabilities is termed a space power (Lupton, 1988, p. 4).

1994—Larned @ Air & Space Doctrine Symposium

Spacepower is the ability to exploit the civil, commercial and national security space systems and associated infrastructure in support of national security strategy (Jusell, 1998, p. 8).

1995—Hyatt et al., Space Power 2010

Space power is the ability of a state **or non-state** actor to achieve its goals and objectives in the presence of other actors on the world stage through control and exploitation of the space environment (Hyatt, Paul, Rampino, Ricchi, & Schwarz, 1995, p. 5).

1996—Gray, The influence of Space Power upon History

Space power may be defined as the ability to use space while denying reliable use to any foe (Gray, 1996, p. 293).

Space power, as mentioned above, refers simply to the ability to use space for military, civil, or commercial purposes and to deny the ability of an enemy to do the same. This functional, output oriented definition (tightly linked to the ability of the nation to control vital spaceways) is preferable to a strictly unilateral, descriptive usage and parallels common definitions of air power and sea power. Any concept which embraces the word "power" accepts the ambiguities of that difficult term. (Gray, 1996, p. 299)

1997—Billman, The Inherent Limitations of Spacepower: Fact or Fiction?

No standard definitions seem to exist for air, land or sea power. However, all seem to have similar characteristics, and hence spacepower can be defined in a similar manner. As Lt Col David E. Lupton writes in his work On Space Warfare:

> "A Spacepower Doctrine: Spacepower is the ability of a nation to exploit the space environment in pursuit of national goals and purposes and includes the entire astronautical capabilities of the nation."

The United States is a spacepower dependent nation. It has a space infrastructure, both civilian and military, and is presently exploiting space for many purposes. As naval forces supply the military component of sea power, and air forces provide the military component of airpower, so too do space forces supply the military component of spacepower.

1998—Air Force Doctrine Document 2-2, Space Operations

In text: Space power is the capability to employ space forces to achieve national security objectives (U.S. Air Force, 1998, p. 1).

Glossary: [Space Power:] The capability to exploit civil, commercial, intelligence, and national security space systems and associated infrastructure to support national security strategy and national objectives from peacetime through combat operations (U.S. Air Force, 1998, p. 32).

1998—Johnson et al., Space: Emerging Options for National Power

Spacepower is connected to other forms of national power such as economic strength, scientific capabilities, and international leadership. [...] Therefore, we would define spacepower as *the pursuit of national objectives through the medium of space and the use of space capabilities.* [...] The effective exercise of spacepower may require, but it is not limited to, the use of military force (Johnson, Pace, & Gabbard, 1998, p. 8).

1999—DeBlois, Beyond the Paths of Heaven: the Emergence of Space Thought

In AFM 1-1, space power is defined as "that portion of aerospace power that exploits the space environment for the enhancement of terrestrial forces and for the projection of combat power to, in, and from space to influence terrestrial conflict." This definition originated in a draft to AFM 2-25 which no longer exists. Another definition is found in the current draft of AFDD 4: "Spacepower is the capability to exploit civil, commercial, intelligence, and national security space systems and associated infrastructure to support national security strategy and national objectives from peacetime through combat operations." This study uses the AFDD 4 definition. Air Force Manual (AFM) 1-1, Basic Aerospace Doctrine of the United States Air Force, vol. 2, March 1992, 300. AFDD 4, "Space Operations Doctrine," draft, 1 May 1995, 3. (Gallegos, "After the Gulf War: Balancing Space Power's Development." In *Beyond the Paths of Heaven: the Emergence of Space Though*, ed. DeBlois, 1999, p. 91)

1999—Oberg, Space Power Theory

Space power is the combination of technology, demographic, economic, industrial, military, national will, and other factors that contribute to the coercive and persuasive ability of a country to politically influence the actions of other states and other kinds of players, or to otherwise achieve national goals through space activity (Oberg, Space Power Theory, 1999, p. 10).

1999—Gray et al., Space Power and the Revolution in Military Affairs: A Glass Half Full?

If space power is defined as the ability in peace, crisis, and war to exert prompt and sustained influence in or from space then the key enabler for space power has to be space control (Gray & Sheldon, Space Power and the Revolution in Military Affairs: A Glass Half Full?, 1999, p. 36).

2001—Lambakis, On the Edge of Earth: The Future of American Space Power

In considering the question of what is a space power, we must look even harder at the evolution, and increasingly international character of, the global commercial infrastructure. Strategic partnerships and private corporations have carried us to the point where national ownership is no longer the only criterion in our definition. It may be more useful, therefore, to regard a space power to be any entity that has the capacity to utilize effectively the space medium for commercial or national security purposes, with some pieces of its space operations coming from dedicated national satellites and others belonging to the private sector and/or government-initiated commercial activities. The baseline measure of space power will be a country's ability to integrate space capabilities with other national activities and manage the rapid and immense flow of information. Clever space powers will be those that can effectively utilize the combinations of all the space services and elements available to it. Superior space powers will own and confidently apply significant space capabilities and possess, as part of their national infrastructures, the requisite skills to exploit them fully (Lambakis, *On the Edge of Earth: The Future of American Space Power*, 2001, p. 46).

2001—Logsdon, Just Say Wait to Space Power

Space power can be defined as using the space medium and assets located in space to enhance and project U.S. military power (Logsdon, 2001).

2001—Smith, Ten Propositions Regarding Spacepower

For the purpose of this study, "*spacepower is defined as the ability of a state or non-state actor to achieve its goals and objectives in the presence of other actors on the world stage through exploitation of the space environment.*"[1] This definition is remarkably similar to a definition for any other form of power, be it air, land, sea, or information. In the broadest sense, spacepower includes all activities performed

[1] This is from Hyatt III et al. (1995).

by an actor—or exploited by an actor—in the space environment for civil, military, commercial, or other reasons. (Smith, 2001, p. 6)

2001—Fox, Some principles of Space Strategy (or "Corbett in Orbit")

"space power"—the ability to use space while denying its reliable use to any foe (Quoting Gray, 1996).

2002—Joint Publication 3–14, Space Operations

The total strength of a nation's capabilities to conduct and influence activities to, in, through, and from space to achieve its objectives (Joint Chiefs of Staff, 2002, p. GL-6).

2004—Moorthi, What 'Space Security' Means for an Emerging Space Power

The term 'spacepower' is used normally with the meaning of 'might'. However, here it is appropriate to describe it as the demonstrated ability to use the power of space for human welfare (Moorthi, p. 261).

2006—Harter, Ten Propositions on Space Power—The Dawn of a Space Force

Space power—a nation's ability to exploit and control the space medium to support and achieve national goals (Harter, 2006, p. 67).

2006—Fredrickson, Globalness—Towards a Space Power Theory

the term *space power* as it is used in this paper refers to military space power and is defined simply as the use of space to achieve military objectives (Fredrickson, p. 4).

2007—Krepon et al., Preserving Freedom of Action in Space: Realizing the Potential and Limits of U.S. Spacepower

Our working definition of spacepower is the sum total of capabilities that contribute to a nation's ability to benefit from the use of space (Krepon et al., p. 1).

2007—Lutes et al., Toward a theory of Spacepower (Presentation)

Spacepower: The ability to use space to get desired outcomes by influencing the environment and the behavior of others (Lutes et al., 2007, p. 14).

2008—Lutes, Spacepower in the 21st Century

Spacepower, then, might be defined as the ability to use space to influence other actors and the external environment to achieve one's objectives. Spacepower both contributes to and is supported by other forms of power: diplomatic, informational, military, and economic, among others. Spacepower can be looked at through socio-cultural, economic, and security lenses, each roughly equating to the civil-scientific, commercial, and military intelligence sectors of space activity (Lutes, Spacepower in the 21st Century, 2008, p. 67).

2008—Peter, Space Power and Europe

Space power can be defined as the "total strength and ability of a State to conduct and influence activities to, in, through and from space to achieve its goals and objectives (security, economic and political), to affect desired outcomes in the presence of other actors in the world stage and if necessary to change the behaviour of others by exploiting the space systems and associated ground-infrastructure as well as political leverage it has garnered" (Peter, 2009).

2010—Peter, Space Power and Its Implications: The Case of Europe

Space power is consequently proposed to be defined as the "total strength and ability of a State to conduct and influence activities to, in, through and from space to achieve its goals and objectives (security and military, economic and political) to affect desired outcomes in the presence of other actors in the world stage and if necessary to change the behaviour of others by exploiting the space systems and associated ground-infrastructure as well as political leverage it has garnered." Space power is therefore the ability to use space to get desired outcomes by influencing the environment and the behaviour of others. In other words, space power is the pursuit of national objectives by the use of space capabilities. It is not a single property, but a combination of factors. Space power is composed of a set of interrelated elements.

2011—Toward a Theory of Spacepower: Selected Essays

An expanded definition of spacepower could then be "the ability to use space to influence others, events, or the environment to achieve one's purposes or goals" (Editors, p. xiv).

Spacepower is defined here as "the ability in peace, crisis, and war to exert prompt and sustained influence in or from space" (Sheldon & Gray p. 2).

Definitions of spacepower focus on the ability, as Colin Gray points out, to use space and to deny its use to enemies. 1. Spacepower is a multifaceted concept that, like power in IR theory, is "complex, indeterminate, and intangible," as Peter L. Hays put it. 2. Spacepower includes the possession of capabilities to conduct military operations in and from space and to utilize space for commercial and other peaceful purposes (Pfaltzgraff, p. 41).

Our working definition of spacepower is the sum total of capabilities that contribute to a nation's ability to benefit from the use of space (Kepron, Hitchens, & Katz-Hyman, 2011, p. 391).

2012—Al-Rodhan, Meta-Geopolitics of Outer Space

I define *space power* as the ability of a state to use space to sustain and enhance its seven state capacities as outlined in the meta-geopolitics framework, namely (social and health, domestic politics, economics, environment, science and human potential, military and security, and international diplomacy). In addition, the governance and sustainability of space power will need to employ a "symbiotic realism" approach to global relations and a "multi-sum security principle" approach to global security.

Ultimately, space will either be safe for everyone or for no one (Al-Rodhan, 2012, p. 25).

2015—Bowen, Spacepower and Space Warfare

Spacepower is the use of space systems and resources towards ultimately political ends (Bowen, p. 80).

2015—Ziarnick, Developing National Power in Space—A Theoretical Model

Space power is the ability to do something in space (Ziarnick, p. 25).

2016—Smith, Spacepower and the Strategist

Spacepower is a vital element of a state's military instrument of power, but spacepower is also a vital element contributing to each instrument of a state's power: diplomacy, information, military, economic and culture (DIME-C) (Smith, p. 160).

2017—Czajkowski, Theory of Spacepower—A Brief Introduction

[T]he capacity to promote the state's interests, to enhance its ability to influence international relations and to maintain national security by the use of orbital systems (Czajkowski, 2017, p. 36).

[…] *spacepower* is an ability of the nation-state to make use of outer space for its own purpose (Czajkowski, 2017, p. 50).

2019—Aliberti et al., Measuring Space Power

[T]his study finds it more useful to regard a space power as an entity with the means to autonomously deploy, operate and benefit from any space-related capability to support the achievement of national objectives (Aliberti, Cappella, & Hrozensky, 2019, p. 15).

2020—Bowen, War in Space: Strategy, Spacepower, Geopolitics

spacepower—a range of space technologies and activities in space—can be deployed and sought by states for the purposes of war, development and prestige (Bowen, 2020, p. 22).

Spacepower is 'the use of outer space's military and economic advantages for strategic ends', and a 'space power' is an entity that uses outer space for its political objectives (Bowen, 2020, p. 22).

Spacepower can be described as both the material capabilities to achieve goals in and from space, as well as the ability to 'use space to influence others, events, or the environment to achieve one's purposes or goals' (Bowen, 2020, p. 23).

2020—Bowen, The Integrated Review and UK Spacepower—The Search for Strategy

Spacepower is 'a diverse collection of activities and technologies in space or to do with outer space… defined by how any actor can use outer space' for the purposes of war, development and prestige (Bowen, 2020, p. 6).

2020—US Space Force, Spacepower—Doctrine for Space Forces

National spacepower is the totality of a nation's ability to exploit the space domain in pursuit of prosperity and security. National spacepower is comparatively assessed as the relative strength of a state's ability to leverage the space domain for diplomatic, informational, military, and economic purposes (USSF, 2020, p. 13).

Appendix C: The "Measuring Spacepower" Survey

Country of Expertise

1. What is your main country of expertise? Please select the country you intend to evaluate in this survey.

    - Argentina
    - Australia
    - Brazil
    - Canada
    - China
    - Europe (EU + ESA and Member States)
    - India
    - Israel
    - Japan
    - Russia
    - Singapore
    - South Africa
    - South Korea
    - United Arab Emirates
    - United States of America

Soft Capacity Questions

Space and Security

2. On a scale from 1 to 4, how well integrated would you say that space is in the national security policies of the country?

    - 1. Not integrated at all
    - 2.
    - 3.
    - 4. Fully integrated

3. On a scale from 1 to 4, how well integrated would you say that the security of space assets is in the national security policies of the country?
    - 1. Not integrated at all
    - 2.
    - 3.
    - 4. Fully integrated
4. On a scale from 1 to 4, how often would you say that space assets are used for surveillance, verification, and/or risk assessment in the country?
    - 1. Never
    - 2.
    - 3.
    - 4. Very often
5. On a scale from 1 to 4, how often would you say that space assets are used in crisis and disaster prevention and/or management in the country?
    - 1. Never
    - 2.
    - 3.
    - 4. Very often

Space and Defence

6. On a scale from 1 to 4, how well integrated would you say that space is in the national military strategy of the country?
    - 1. Not integrated at all
    - 2.
    - 3.
    - 4. Very integrated
7. On a scale from 1 to 4, how often would you say that the country uses space for the prevention and/or deterrence of hostile actions?
    - 1. Never
    - 2.
    - 3.
    - 4. Very often
8. On a scale from 1 to 4, how often would you say that the country uses space in military operations when it comes to command, control, communications, computing (C4)?
    - 1. Never
    - 2.
    - 3.
    - 4. Very often

9. On a scale from 1 to 4, how often would you say that the country uses space in military operations when it comes to military intelligence, surveillance, and reconnaissance capabilities (ISR), including early warning, signal interception, and active observation?
   - 1. Never
   - 2.
   - 3.
   - 4. Very often
10. On a scale from 1 to 4, how often would you say that the country uses space in military operations when it comes to other military support services (e.g., augmentation of terrestrial technologies as weather forecasting, data transfer, logistical support, missile guidance, etc.)?
    - 1. Never
    - 2.
    - 3.
    - 4. Very often

Space and Foreign Policy

11. On a scale from 1 to 4, how often would you say that the country uses space for diplomatic purposes, be them political, strategic, and/or economic?
    - 1. Never
    - 2.
    - 3.
    - 4. Very often
12. On a scale from 1 to 4, how influential would you say that the country is in international space fora (e.g., COPUOS, CD, etc.)?
    - 1. Not influential at all
    - 2.
    - 3.
    - 4. Very influential
13. On a scale from 1 to 4, how often would you say that the country uses space to support foreign aid and/or international initiatives (e.g., UN SDGs)?
    - 1. Never
    - 2.
    - 3.
    - 4. Very often

14. On a scale from 1 to 4, how successful would you say that the country is in using space to create/boost its international prestige?

    - 1. Not successful at all
    - 2.
    - 3.
    - 4. Very successful

Environment and Resources

15. On a scale from 1 to 4, how often would you say that space assets are used to support the agricultural sector in the country?

    - 1. Never
    - 2.
    - 3.
    - 4. Very often

16. On a scale from 1 to 4, how often would you say that space assets are used to support meteorology and weather forecasting activities in the country?

    - 1. Never
    - 2.
    - 3.
    - 4. Very often

17. On a scale from 1 to 4, how often would you say that space assets are used to support natural resources management (e.g., forestry, fishing, mining, etc.) in the country?

    - 1. Never
    - 2.
    - 3.
    - 4. Very often

18. On a scale from 1 to 4, how often would you say that space assets are used to support environmental monitoring and/or protection (biodiversity and ecosystems) in the country?

    - 1. Never
    - 2.
    - 3.
    - 4. Very often

19. On a scale from 1 to 4, how often would you say that space assets are used to support climate change monitoring and/or mitigation policies in the country?
    - 1. Never
    - 2.
    - 3.
    - 4. Very often

Infrastructure

20. On a scale from 1 to 4, how often would you say that space assets are used to support infrastructure management (e.g., construction, logistics, finance, etc.) in the country?
    - 1. Never
    - 2.
    - 3.
    - 4. Very often
21. On a scale from 1 to 4, how often would you say that space assets are used to support the energy sector (e.g., site identification, pipelines and grids timing and synchronization, etc.) in the country?
    - 1. Never
    - 2.
    - 3.
    - 4. Very often
22. On a scale from 1 to 4, how often would you say that space assets are used to support transport and/or mobility (e.g., land, water, air navigation, traffic monitoring, goods tracking, etc.) in the country?
    - 1. Never
    - 2.
    - 3.
    - 4. Very often

Development and Growth

23. On a scale from 1 to 4, how often would you say that space assets are used to support urban and/or rural development (e.g., survey and mapping, development plans, wasteland management, etc.) in the country?
    - 1. Never
    - 2.
    - 3.
    - 4. Very often

24. On a scale from 1 to 4, how often would you say that space activities contribute to scientific and/or technological innovation in the country?

- 1. Never
- 2.
- 3.
- 4. Very often

25. On a scale from 1 to 4, how often would you say that space activities contribute to the development of the industrial base in the country?

- 1. Never
- 2.
- 3.
- 4. Very often

26. On a scale from 1 to 4, how successful would you say that the country is in using space to stimulate market development and commercial activities?

- 1. Not successful at all
- 2.
- 3.
- 4. Very successful

Civil Society

27. On a scale from 1 to 4, how often would you say that space assets are used to support the education sector (e.g., remote learning services) in the country?

- 1. Never
- 2.
- 3.
- 4. Very often

28. On a scale from 1 to 4, how often would you say that space assets are used to support the health sector (e.g., telemedicine services) in the country?

- 1. Never
- 2.
- 3.
- 4. Very often

29. On a scale from 1 to 4, how often would you say that space assets are used to provide entertainment and other citizen services (e.g., broadcasting, internet services, GIS, etc.) in the country?

- 1. Never
- 2.
- 3.
- 4. Very often

30. On a scale from 1 to 4, how successful would you say that the country is in using space to create/boost national identity and social cohesion?

    - 1. Not successful at all
    - 2.
    - 3.
    - 4. Very successful

Political Autonomy Questions

31. On a scale from 1 to 4, how autonomous from foreign nations would you say that the country is when it comes to the decision to join space-related bilateral and/or multilateral arrangements?

    - 1. Not autonomous at all
    - 2.
    - 3.
    - 4. Fully autonomous

32. On a scale from 1 to 4, how autonomous from the national military would you say that the country is when it comes to the decision to join space-related bilateral and/or multilateral arrangements?

    - 1. Not autonomous at all
    - 2.
    - 3.
    - 4. Fully autonomous

33. On a scale from 1 to 4, how autonomous from domestic corporations (state or private) would you say that the country is when it comes to the decision to join space-related bilateral and/or multilateral arrangements?

    - 1. Not autonomous at all
    - 2.
    - 3.
    - 4. Fully autonomous

Acting

34. On a scale from 1 to 4, how autonomous from foreign nations would you say that the country is when it comes to acting (e.g., voting, coalition building, etc.) within major international space fora?

    - 1. Not autonomous at all
    - 2.
    - 3.
    - 4. Fully autonomous

35. On a scale from 1 to 4, how autonomous from the national military would you say that the country is when it comes to acting (e.g., voting, coalition building, etc.) within major international space fora?

- 1. Not autonomous at all
- 2.
- 3.
- 4. Fully autonomous

36. On a scale from 1 to 4, how autonomous from domestic corporations (state or private) would you say that the country is when it comes to acting (e.g., voting, coalition building, etc.) within major international space fora?

- 1. Not autonomous at all
- 2.
- 3.
- 4. Fully autonomous

Complying

37. On a scale from 1 to 4, how autonomous from foreign nations would you say that the country is when it comes to complying with space-related international law (including both 'hard' and 'soft' law)?

- 1. Not autonomous at all
- 2.
- 3.
- 4. Fully autonomous

38. On a scale from 1 to 4, how autonomous from the national military would you say that the country is when it comes to complying with space-related international law (including both 'hard' and 'soft' law)?

- 1. Not autonomous at all
- 2.
- 3.
- 4. Fully autonomous

39. On a scale from 1 to 4, how autonomous from domestic corporations (state or private) would you say that the country is when it comes to complying with space-related international law (including both 'hard' and 'soft' law)?

- 1. Not autonomous at all
- 2.
- 3.
- 4. Fully autonomous

National Policies

40. On a scale from 1 to 4, how autonomous from foreign nations would you say that the country is when it comes to formulating a 'national space policy'?
    - 1. Not autonomous at all
    - 2.
    - 3.
    - 4. Fully autonomous
41. On a scale from 1 to 4, how autonomous from the national military would you say that the country is when it comes to formulating a 'national space policy'?
    - 1. Not autonomous at all
    - 2.
    - 3.
    - 4. Fully autonomous
42. On a scale from 1 to 4, how autonomous from domestic corporations (state or private) would you say that the country is when it comes to formulating a 'national space policy'?
    - 1. Not autonomous at all
    - 2.
    - 3.
    - 4. Fully autonomous

Programmes

43. On a scale from 1 to 4, how autonomous from foreign nations would you say that the country is when it comes to defining a 'national space programme'?
    - 1. Not autonomous at all
    - 2.
    - 3.
    - 4. Fully autonomous
44. On a scale from 1 to 4, how autonomous from the national military would you say that the country is when it comes to defining a 'national space programme'?
    - 1. Not autonomous at all
    - 2.
    - 3.
    - 4. Fully autonomous

45. On a scale from 1 to 4, how autonomous from domestic corporations (state or private) would you say that the country is when it comes to defining a 'national space programme'?
    - 1. Not autonomous at all
    - 2.
    - 3.
    - 4. Fully autonomous

Partners

46. On a scale from 1 to 4, how autonomous from foreign nations would you say that the country is when it comes to choosing its partners within the space domain?
    - 1. Not autonomous at all
    - 2.
    - 3.
    - 4. Fully autonomous
47. On a scale from 1 to 4, how autonomous from the national military would you say that the country is when it comes to choosing its partners within the space domain?
    - 1. Not autonomous at all
    - 2.
    - 3.
    - 4. Fully autonomous
48. On a scale from 1 to 4, how autonomous from domestic corporations (state or private) would you say that the country is when it comes to choosing its partners within the space domain?
    - 1. Not autonomous at all
    - 2.
    - 3.
    - 4. Fully autonomous

Dominance

49. On a scale from 1 to 4, how often would you say the country attempts to influence, coerce, or buy other countries' decision-making on space-related issues?
    - 1. Never
    - 2.
    - 3.
    - 4. Very often

50. On a scale from 1 to 4, how successful would you say that domestic corporations (state or private) are when it comes to influencing foreign nations' domestic decision-making on space-related issues?

    - 1. Not successful at all
    - 2.
    - 3.
    - 4. Very successful

51. On a scale from 1 to 4, how often would you say that the country expresses a preference for local over foreign solutions when it comes to space technology, infrastructure, and/or manufacture?

    - 1. Never
    - 2.
    - 3.
    - 4. Very often

MIX
Papier aus verantwortungsvollen Quellen
Paper from responsible sources
FSC® C105338

If you have any concerns about our products,
you can contact us on
ProductSafety@springernature.com

In case Publisher is established outside the EU,
the EU authorized representative is:
Springer Nature Customer Service Center GmbH
Europaplatz 3, 69115 Heidelberg, Germany

Printed by Libri Plureos GmbH
in Hamburg, Germany